TANYA ZAKOWICH, CREATOR OF PINK PENCIL MATH

50 MATH TRICKS

THAT WILL CHANGE YOUR LIFE

MENTALLY SOLVE THE IMPOSSIBLE IN SECONDS

PAGE STREET
PUBLISHING CO.

First published in 2023 by
Page Street Publishing Co.
27 Congress Street, Suite 1511
Salem, MA 01970
www.pagestreetpublishing.com

Distributed by Macmillan, sales in Canada by The Canadian Manda Group.

27 26 25 24 2 3 4 5

ISBN-13: 978-1-64567-828-1

ISBN-10: 1-64567-828-8

Library of Congress Control Number: 2022952242

Cover and book design by Molly Kate Young for Page Street Publishing Co.

Illustrations by Tanya Zakowich

Printed and bound in the United States of America

Page Street Publishing protects our planet by donating to nonprofits like The Trustees, which focuses on local land conservation.

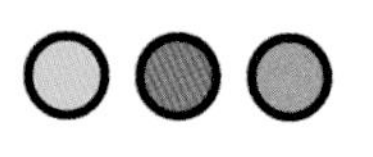

To Ming, Paul, Fiona and Amanda—thank you for your love and support.

Contents

Introduction

When I was twelve, I had to repeat a year of math.

I couldn't wrap my head around the concepts and the teacher didn't think I was ready enough to move on to the next level. Why is it impossible to divide a number by zero? Why is the *order of operations* in that order? I had no idea, that's just how things were. Math seemed like a bunch of arbitrary rules and procedures and the long lectures, textbook definitions and repetitive worksheets didn't click for me.

And then I got hit with the ultimate student nightmare: repeating an entire year of math. But as it turned out, this was the best thing that could have happened to me. With a new teacher came a fresh perspective on old problems, and I realized that there's no one perfect way to do math.

Just as clothing styles vary, so do our approaches to math problem-solving. Some people prefer breaking down problems into small steps, while others tackle the big picture or estimate first. Some people even like to work backward, guess and check or draw pictures.

That year, I developed a style that worked best for me: a dual approach. I'd find two different ways to solve the same problem, which took more time but helped me feel more creative and confident that I was getting the right answer. This style stuck with me throughout college and my career as a mechanical engineer for projects at NASA, Boeing® and Hyperloop One.

After experimenting with various problem-solving techniques over the years, I started a TikTok channel, Pink Pencil Math, to share my favorite tips and tricks. Much to my surprise, my experience with math was far from unique. Millions of people around the world watched the videos to learn different approaches to and perspectives on solving math problems. And the best part? I kept learning new ways to work with numbers too, thanks to all the comments and messages from my fellow viewers!

In this book, I'll walk you through fifty of my all-time favorite mental math tricks. But we won't stop there. We'll also dive deep into the tricks, uncovering the "why" behind each one and exploring how you can apply them in your everyday life. Forget what you thought you knew about math: It's not a rigid set of rules, but a dynamic, creative game that can be approached from all sorts of angles. So, find a comfortable chair, grab your favorite drink and let's unleash your inner mathematician!

Tanya Zakowich

What Is a Math Trick?

This is no ordinary math book.

It holds within its pages a knowledge that is only known to a few.

Think of every math problem you come across in life as a riddle. It begins with a question, it ends in an answer and it's up to you to pick a path to get there. Sure, you might have learned specific ways to solve specific problems, but who says there aren't other ways? There could be many paths to solve a math problem—some could be long, others short and some so short it even looks like you're doing magic.

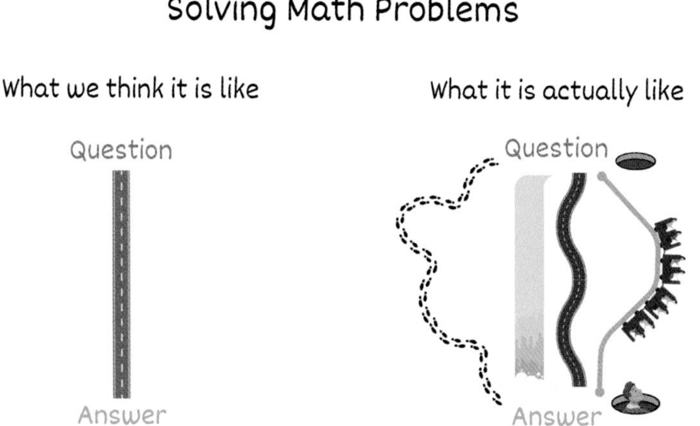

This book is all about finding the quickest and most innovative ways to solve math problems. It will stretch your creative mind like a rubber band until you are able to solve even the most challenging problems in your head! To others, it may look like a math trick, but you'll simply be breaking down numbers in a way that is not commonly understood.

5 X 18

Take a look at this multiplication problem **5 x 18**. Can you solve it in your head? One thing I like to do whenever I multiply two numbers together is to visualize the amounts as two sides of a rectangle. When you multiply the two sides together, you get the "area" of the rectangle, which is also the solution to the original multiplication problem. If you ever feel overwhelmed by a math problem, try visualizing how the numbers can relate to each other. It can go a long way!

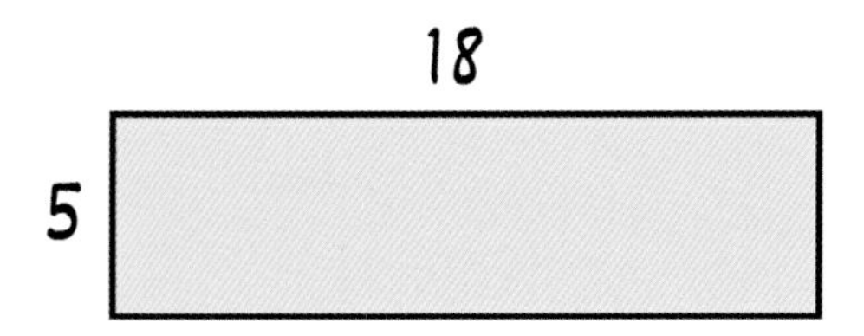

So, how did you go about multiplying **5 x 18** in your head? Here are six possible ways you can solve it. Did you use one of these ways or do something completely different?

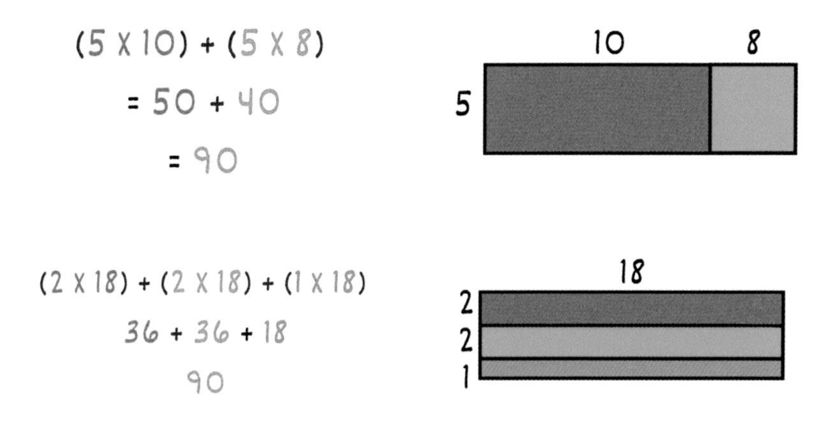

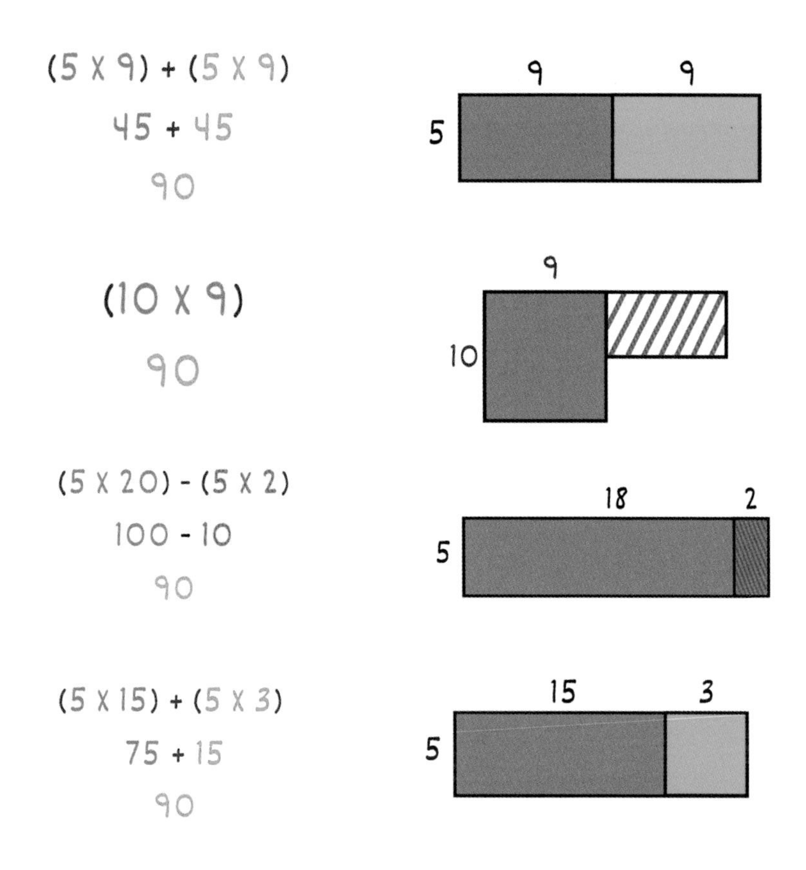

It's much easier to mentally calculate **5 x 18** when you break those numbers down, isn't it? The real challenge is discovering the best way that works for you and getting to that "aha!" moment. But don't worry, this book will help you exercise your creativity and make it all easier. Ready for another example? Let's see if you can figure out why this next calculation works!

Take off your shoe and check your shoe size.

Now, add two zeros to the end of your shoe size.

800

Next, subtract the year you were born.

800 - 1992

Finally, add the current year. The last two digits of your final number will reveal your age (or how old you will be turning this year). Give this a try!

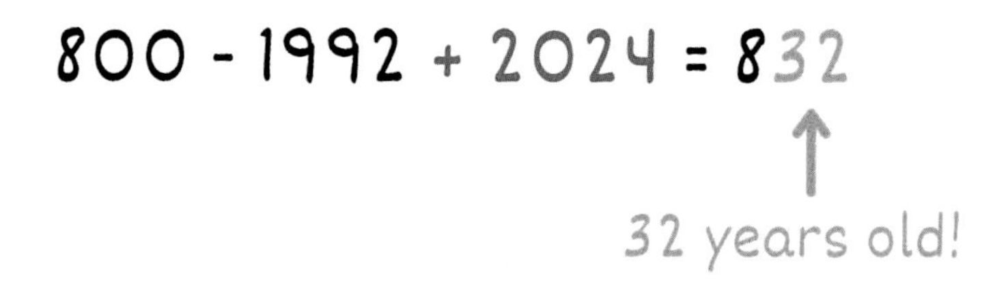

How did we get your age from your shoe size? Well, whenever you subtract the year you were born from the current year, you'll get your age. Adding your shoe size is just a fun distraction from the simple calculation. When you add two zeros to your shoe size, you're shifting your shoe size to the hundreds place and beyond, while your birthday will remain the same in the ones and tens place. Here's what your answer will look like for shoe sizes 6, 8 and 11.

600 - 1992 + 2024 = 632

800 - 1992 + 2024 = 832

1100 - 1992 + 2024 = 1132

So, how do we train our minds to find the easiest path to solve problems? It all boils down to a little creativity using our number sense! We all have five senses—sight, smell, taste, touch and hearing—but what you may not know is that we are also born with a sixth sense: number sense.

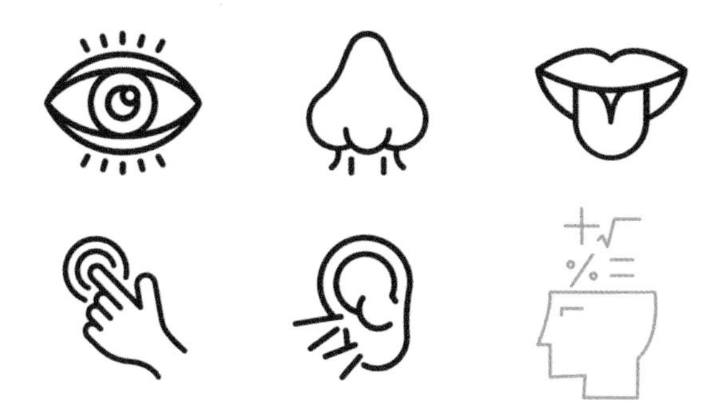

This special ability helps us estimate quantities, understand patterns and make comparisons. Let me show you how it works in action. Take a quick glance at the dots below. How many dots do you see?

I bet you knew in an instant that there are three dots on the page without having to go through and individually count each one, right? How about these dots? How many are there this time?

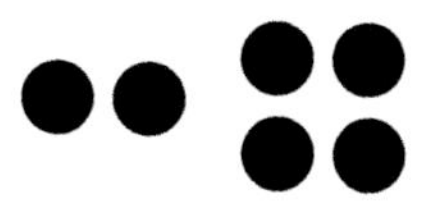

You intuitively knew that there were two dots on the left and another four on the right that make up six total dots. This is using your number sense without even knowing it! It's amazing how our brain can process and understand numbers quickly and accurately by subconsciously recognizing patterns. And not only that, our number sense also helps us estimate numbers in a way that makes sense to us.

For example, how long do you think a million seconds is?

A million seconds is about 12 days! Okay, how about a billion seconds this time?

A billion seconds is a staggering 32 years!

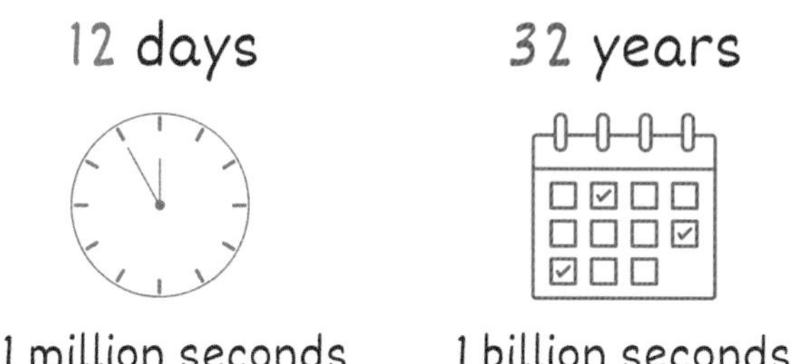

If you thought a billion seconds would be a shorter amount of time, don't worry, most of us do. A billion is 1,000 times bigger than a million, but it's difficult for our minds to visualize that difference without exercising our number sense. Check out these two piles of rice where each grain represents 100,000. The pile on the left makes a million and the pile on the right makes a billion! Can you believe how much more rice there is in that pile?

By training your number sense throughout this book, you'll learn how to better estimate quantities, shape-shift numbers and discover the shortest paths to solve even the most difficult math problems. This book is your key to unlocking a whole new way of approaching math and it's going to be a fun journey. So, are you ready to get creative and start unlocking the secrets of math together? Let's begin!

The Magic of 0, 1, 2, 10 and 100

Are you ready to learn the trick behind a math trick? Here's the secret—all you're doing is breaking down difficult numbers into smaller ones that you can easily work with. That's it! Let me show you what I mean, but first, what makes a number *difficult* in the first place?

When we think of a difficult number, many of us picture a big number like 9,395,872. Big numbers can be difficult, but it's not always about size. The type of number and how it's presented can also make it tricky. For example, take **20 x 300**. At first glance, this calculation may look daunting, but it's much easier to solve than **12 x 23**. That's because 20 is just **2 x 10** and 300 is just **3 x 100**, so you can simply multiply **2 x 3** to get 6 and then multiply it by 1,000 to get 6,000.

This brings me to our magic numbers: 0, 1, 2, 10 and 100.

These numbers will become your new best friends because they are by far the easiest numbers your brain can work with. They can make even the most daunting calculations seem like a walk in the park. But why these numbers? Let's take a closer look.

Multiplying anything by 2 is simple because it's easy for our brains to double any number, even big ones! Try **34 x 2** and **132 x 2**. You should be able to easily multiply these two to get 68 and 264!

Multiplying anything by 10 is also easy; simply shift the decimal point once to the right (**7 x 10 = 70**). Dividing by 10? Shift the decimal point once to the left (**7 ÷ 10 = 0.7**) Multiplying or dividing by 100? Simply shift the decimal point twice to the right or left. And for 1,000? Shift it three times. It's that simple!

7 X 10 → 7.0 → 70

7 X 100 → 7.0 → 700

7 X 1000 → 7.0 → 7000

7 ÷ 10 → 7.0 → 0.7

7 ÷ 100 → 7.0 → 0.07

7 ÷ 1000 → 7.0 → 0.007

But what if your problem doesn't have these magic numbers in it? The good news is you can break any number down into one or more of our magic numbers. And it doesn't stop there; numbers with trailing zeros such as 20, 30, 40, 500, 600, 7,000 and 80,000 are also great numbers to work with.

Take the example from earlier, **12 x 23**. An easy way to solve this problem is by breaking 12 down into **10 + 2** and multiplying them by 23, but there are many other ways as well!

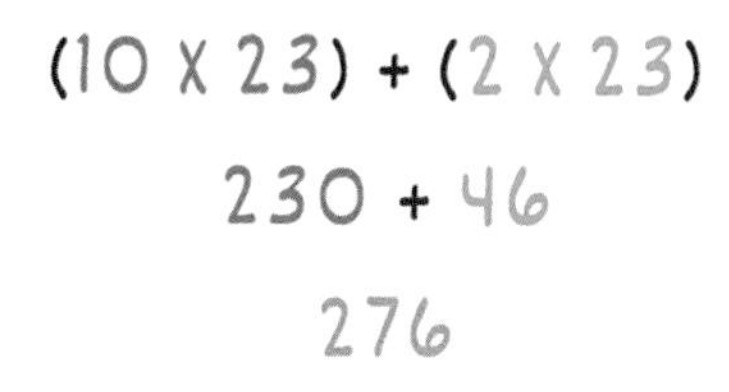

Remember when we previously visualized multiplying two numbers as multiplying two sides of a rectangle to find the area? We can do this again for **12 x 23**!

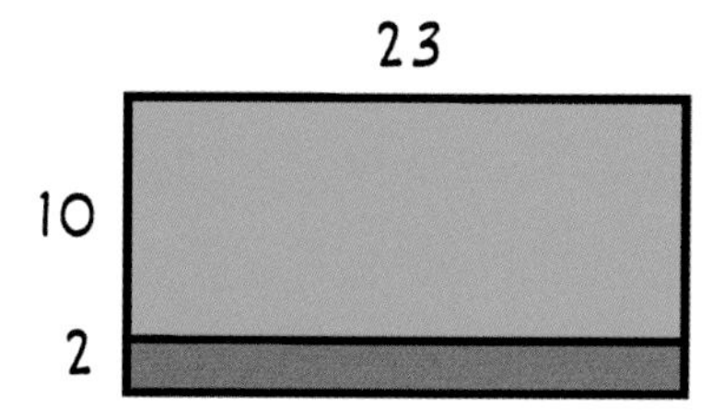

By learning to recognize and work with "magic numbers," you will soon master the art of mental math and be able to solve the most difficult problems in seconds. So, start breaking down your numbers and be amazed at what you can do without a calculator!

Never Forget Your Times Tables

Welcome to the first step in an exciting journey to master mental math! It all starts with the foundation, your times tables from 2 to 10.

x	1	2	3	4	5	6	7	8	9	10
1	1	2	3	4	5	6	7	8	9	10
2	2	4	6	8	10	12	14	16	18	20
3	3	6	9	12	15	18	21	24	27	30
4	4	8	12	16	20	24	28	32	36	40
5	5	10	15	20	25	30	35	40	45	50
6	6	12	18	24	30	36	42	48	54	60
7	7	14	21	28	35	42	49	56	63	70
8	8	16	24	32	40	48	56	64	72	80
9	9	18	27	36	45	54	63	72	81	90
10	10	20	30	40	50	60	70	80	90	100

With these times tables under your belt, you'll be able to use them to mentally calculate bigger numbers like **13 x 17**. That's right, soon you'll be able to instantly spit out the answer for **13 x 17** without ever having to memorize the 13 or 17 times tables.

If you're tired of memorizing multiplication charts, you'll love this section. We're going to shake things up by rearranging numbers into shapes and searching charts for patterns within them. Let's begin!

0, 2, 4, 6, 8, Repeat

Let's start all the way at the beginning with the 2 times table. I'm sure you're very familiar with this times table, but have you ever written it out using a grid?

Steps

Draw a grid with two rows and five columns.

2 Next, write 0, 2, 4, 6, 8 across each row. These will be in the ones place.

0	2	4	6	8
0	2	4	6	8

Now write 0 across the first row. These will be in the tens place.

00	02	04	06	08
0	2	4	6	8

What comes after 0? Well, 1 does! Write 1 in the tens place across in the second row.

00	02	04	06	08
10	12	14	16	18

You've now completed your 2 times table from 2 × 0 to 2 × 9!

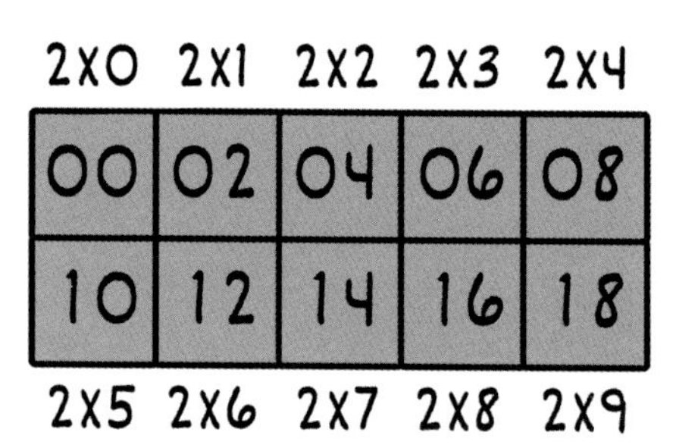

Bonus

Want to keep going beyond **2 x 9**? Just follow this simple pattern: write 0, 2, 4, 6, 8 in the ones place and increase your tens place by adding a 1 every row down. Check out the grid below. How would you fill out the next few rows?

00	02	04	06	08
10	12	14	16	18

3 in a Row, Tic-Tac-Toe

The 2 times table was just a warm-up; the 3 times table is where the real excitement begins! Remember the game Tic-Tac-Toe? Well, here's how you can use it for the 3 times table for **3 x 1** to **3 x 9**.

Steps

 Draw a Tic-Tac-Toe grid.

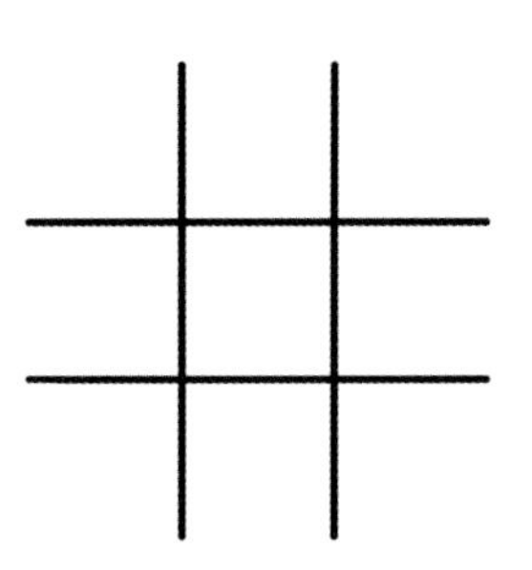

2. Count from 1 to 9 going up each column from left to right. These numbers will be in the ones place for your 3 times table.

3	6	9
2	5	8
1	4	7

↑

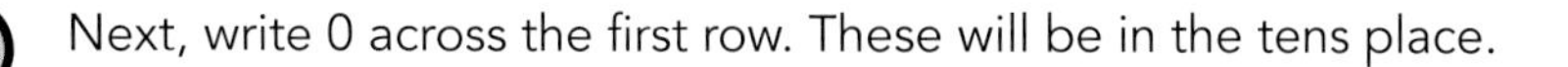

Next, write 0 across the first row. These will be in the tens place.

03	06	09
2	5	8
1	4	7

What comes after 0? Well, 1 does! Write 1 in the tens place across in the second row.

03	06	09
12	15	18
1	4	7

The next number after 1 is 2. Write 2 in the tens place across the next row.

03	06	09
12	15	18
21	24	27

And with that, you have your 3 times table from **3 x 1** to **3 x 9**!

3x1 03	3x2 06	3x3 09
3x4 12	3x5 15	3x6 18
3x7 21	3x8 24	3x9 27

Can You Draw 6 Xs?

Before we move on, let's take a quick break from all the math and have some fun! Here is my favorite logic puzzle that uses a Tic-Tac-Toe grid. It's a great way to exercise your brain while taking a little breather. Here we go!

Can you draw 6 Xs in this grid without having 3 Xs in a row?

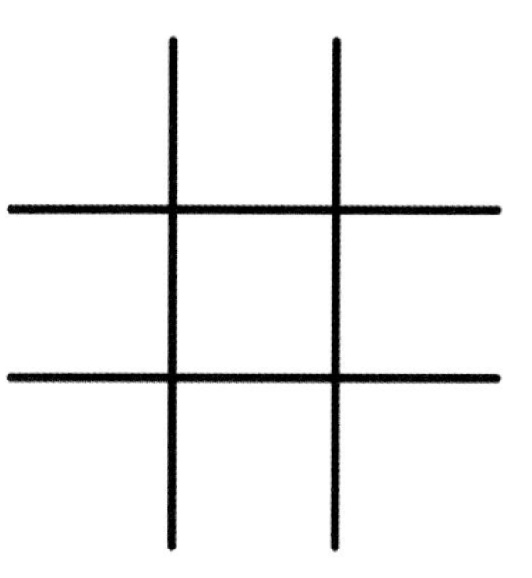

Give it a try and then check your answer at the back of the book!

4 Layers of Bricks

Have you ever taken a close look at a brick wall? It's not just a bunch of bricks stacked on top of each other like building blocks. Instead, the bricks are arranged in an alternating pattern.

This layout not only creates a strong wall, but also reduces the chance of the bricks cracking. And guess what? We're going to use this same brick-laying pattern to remember our 4 times table. So, put your hard hat on and let's build our 4 times table from 4 x 0 to 4 x 9!

Steps

Use this brick-laying pattern to layer your first two rows of bricks. Place three bricks on top and two bricks on the bottom.

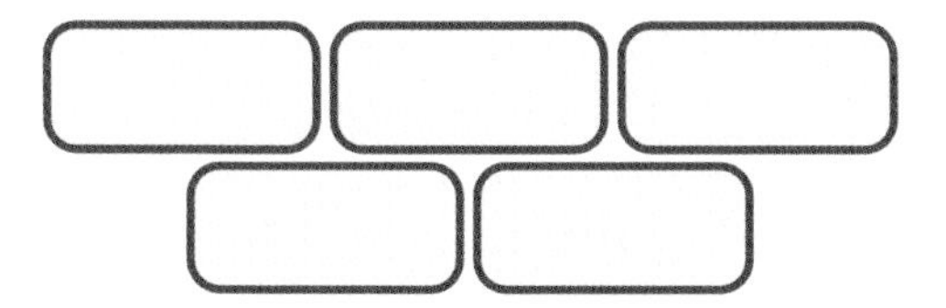

2 Write 0, 2, 4, 6, 8 in a zigzag pattern from left to right. These numbers will be in the ones place.

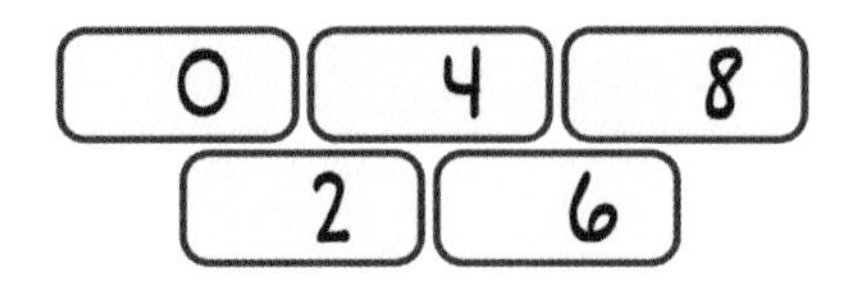

Now add another two rows of bricks below using the brick-laying pattern.

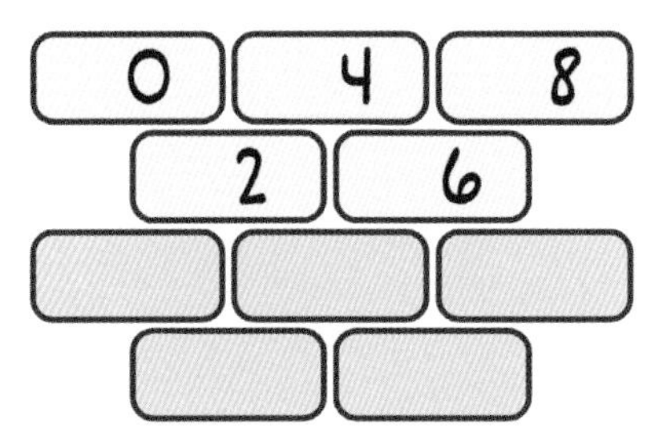

Write 0, 2, 4, 6, 8 again using the same zigzag pattern as earlier.

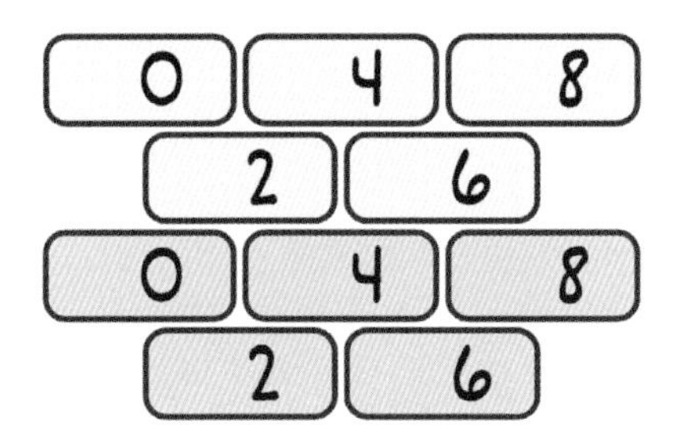

With all the ones digits in place, you're now ready to fill in the tens digits! These will always be 1 less than the row number they're in. For example, write 0 across the first row, 1 across the second row, 2 across the third row and 3 across the fourth row.

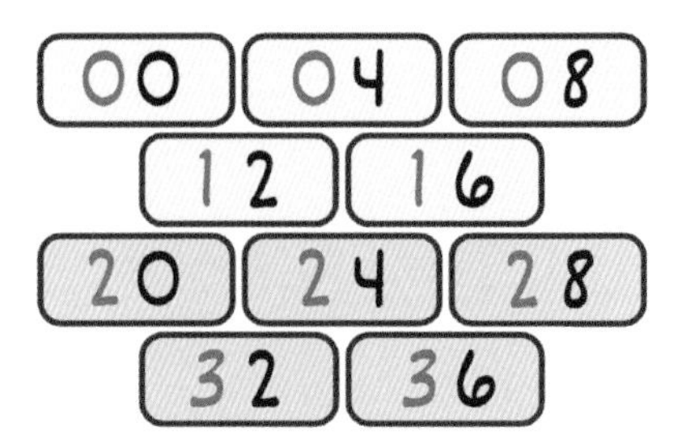

6 You have now completed your 4 times table from **4 x 0** to **4 x 9**. Nice!

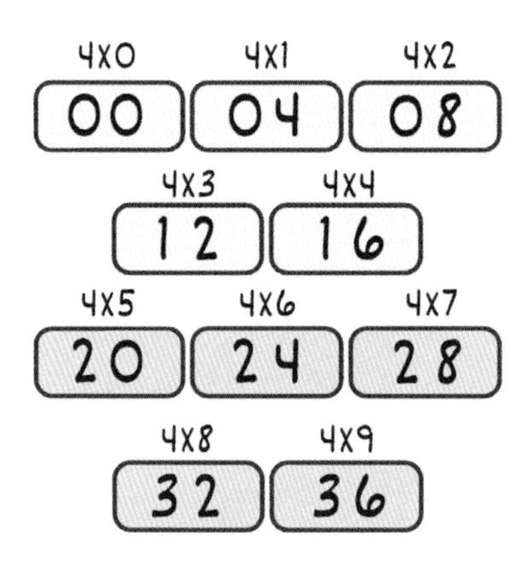

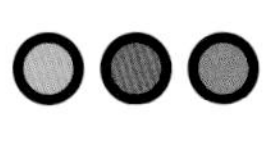

Bonus

How would you continue this brick pattern to find **4 x 10** and beyond? All you need to do is add another two rows of bricks, write 0, 2, 4, 6, 8 in the ones place in a zigzag pattern and fill the tens place in ascending order. Here is **4 x 0** to **4 x 14**, but you can keep doing this to get **4 x 15** and beyond!

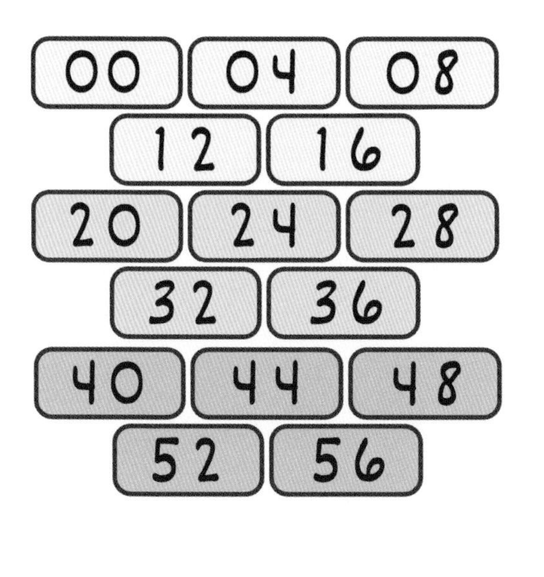

Seesaw Between 5 and 0

Let's whip out our 5 times table chart because I want to show you a pattern. I call it *Seesaw Between 5 and 0* because all your numbers will end in a 5 or 0!

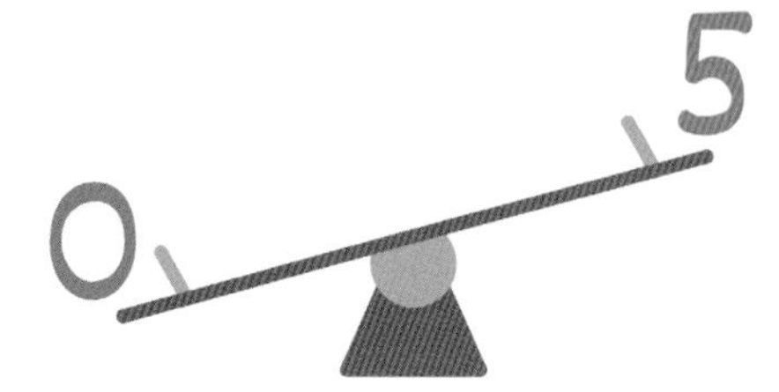

Steps

Write 0 and 5 alternating all the way down from **5 x 0** to **5 x 13**. These will be in the ones place.

0 X 5 = 0
1 X 5 = 5
2 X 5 = 0
3 X 5 = 5
4 X 5 = 0
5 X 5 = 5
6 X 5 = 0
7 X 5 = 5
8 X 5 = 0
9 X 5 = 5
10 X 5 = 0
11 X 5 = 5
12 X 5 = 0
13 X 5 = 5

Write 0 to 6 twice in the tens place (0, 0, 1, 1, 2, 2 . . . 5, 5, 6, 6). With that, you have completed the 5 times table from **5 x 0** to **5 x 13**. But wait, there's more! You can keep repeating this pattern to get **5 x 14** and beyond.

0 X 5 = 00
1 X 5 = 05
2 X 5 = 10
3 X 5 = 15
4 X 5 = 20
5 X 5 = 25
6 X 5 = 30
7 X 5 = 35
8 X 5 = 40
9 X 5 = 45
10 X 5 = 50
11 X 5 = 55
12 X 5 = 60
13 X 5 = 65

Double Tic-Tac-Toe for 6

Ready to take Tic-Tac-Toe to the next level? This time let's use two grids to create the 6 times table from 6 x 1 to 6 x 10!

Steps

Draw two Tic-Tac-Toe grids side by side.

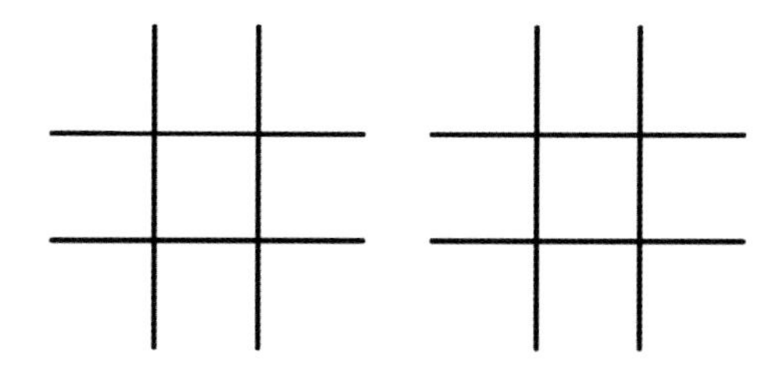

2 Write a 0 below the first column in each grid. Then, count to 9 going up each column from left to right. These numbers will be in the ones place for your 6 times table.

3	6	9		3	6	9
2	5	8		2	5	8
1	4	7		1	4	7
0				0		

How about the numbers in the tens place? These will be the same across each row and increase by 1 down each row. Starting with the Tic-Tac-Toe grid on the left, write 0 across the first row, 1 across the second row, 2 across the third row and finally 3 at the bottom.

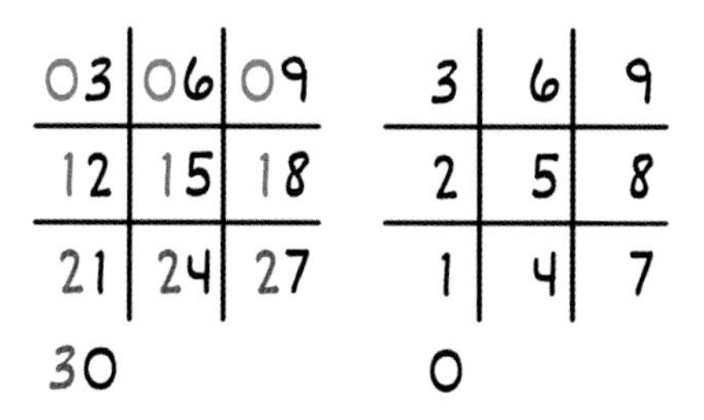

To fill the tens digits for the right grid, start at 3 across the first row and work your way down each row to 6.

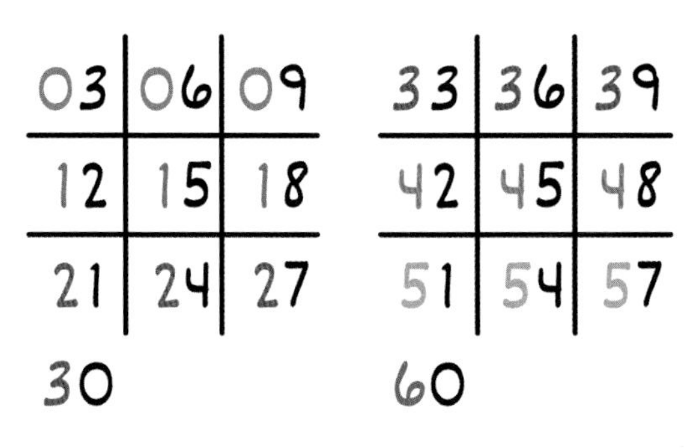

Now comes the interesting part! Draw a line in the shape of a sideways baseball cap on each grid. The lines should pass through five numbers.

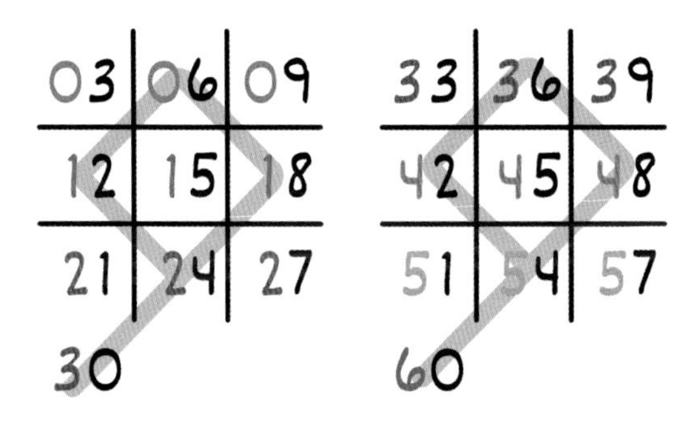

6

The 6 times table will include every number on the baseball cap lines. Read them in order from top to bottom starting with the Tic-Tac-Toe grid on the left!

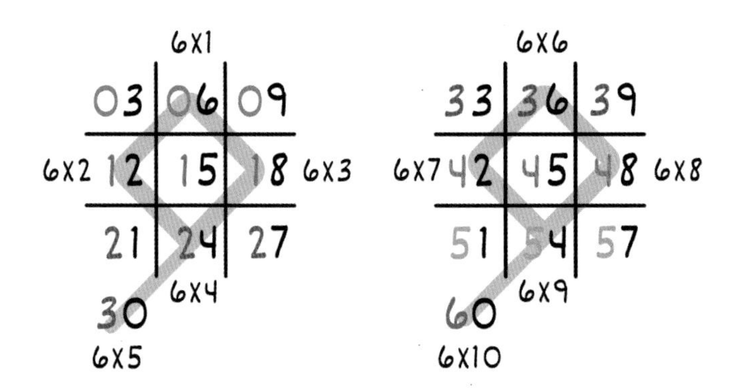

7 in a Row, Tic-Tac-Toe

Are you ready to tackle the 7 times table? So far, we've used the Tic-Tac-Toe grid for the 3 and 6 times tables, but let's put it to use one last time. This will be much easier than when we used two grids for the 6 times table, I promise!

Draw one Tic-Tac-Toe grid.

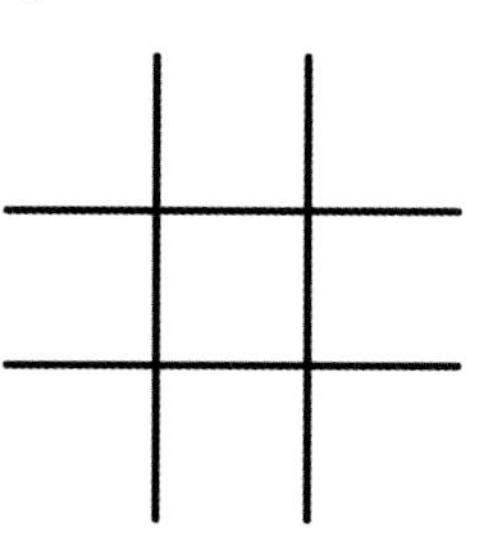

Write 1 to 9 starting at the top right and ending at the bottom left. These numbers will be in the ones place.

7	4	1
8	5	2
9	6	3

Write 0, 1, 2 across the first row. These ascending numbers will be in the tens place.

07	14	21
8	5	2
9	6	3

The tens digits in the second row will also be in ascending order, but this time starting with 2.

07	14	21
28	35	42
9	6	3

Finally, start the ascending-ordered tens digits in the third row with 4.

07	14	21
28	35	42
49	56	63

6. You've now completed your 7 times table from **7 x 1** to **7 x 9**! They are ordered from left to right across each row.

7X1 07	7X2 14	7X3 21
7X4 28	7X5 35	7X6 42
7X7 49	7X8 56	7X9 63

Fractions Over 7? Remember 142857

Speaking of the number 7, I want to show you something fascinating that happens with 7 before we continue on with our times tables!

Whenever you have a fraction over 7, it'll always convert into a long repeating decimal. But here's the cool part—there's a hidden pattern in each decimal! If you look closely, you'll see the sequence 142857. All proper fractions over 7 will repeat this sequence over and over again, but in a different order!

142857

$\frac{1}{7} = 0.1428571428...$

$\frac{2}{7} = 0.2857142857...$

$\frac{3}{7} = 0.4285714285...$

$\frac{4}{7} = 0.5714285714...$

$\frac{5}{7} = 0.7142857142...$

$\frac{6}{7} = 0.8571428571...$

Cross the River for 8

Are you finding it hard to memorize your 8 times table? Try using this simple pattern and you'll be an 8-times-table master in no time. Just remember—cross the river down the middle!

Steps

1. Let's whip out the 8 times table chart. The first thing you're going to do is draw a line between **5 x 8** and **6 x 8**. I like to think of this as a river we'll be crossing later on.

1 X 8 =
2 X 8 =
3 X 8 =
4 X 8 =
5 X 8 =
~~~~~~~~~~
6 X 8 =
7 X 8 =
8 X 8 =
9 X 8 =
10 X 8 =
~~~~~~~~~~

2 Let's first start with the tens digits. Write 0 to 4 going down from the top to the river.

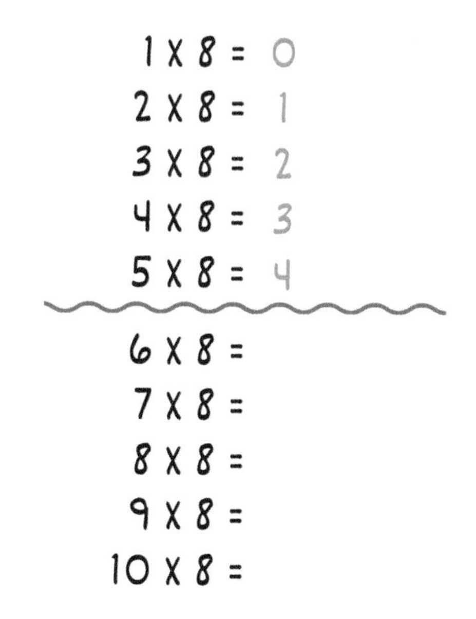

Cross the river and continue down, writing 4 to 8.

1 X 8 = 0
2 X 8 = 1
3 X 8 = 2
4 X 8 = 3
5 X 8 = 4

~~~~~~~~~~

6 X 8 = 4
7 X 8 = 5
8 X 8 = 6
9 X 8 = 7
10 X 8 = 8
~~~~~~~~~~

4. Now go back up to the river and fill 0, 2, 4, 6, 8 in the ones place.

1 X 8 = 0
2 X 8 = 1
3 X 8 = 2
4 X 8 = 3
5 X 8 = 4

6 X 8 = 48
7 X 8 = 56
8 X 8 = 64
9 X 8 = 72
10 X 8 = 80

5. Cross the river again and continue up with another 0, 2, 4, 6, 8 and voilà! You've completed your 8 times table from **8 x 1** to **8 x 10**!

1 X 8 = 08
2 X 8 = 16
3 X 8 = 24
4 X 8 = 32
5 X 8 = 40

6 X 8 = 48
7 X 8 = 56
8 X 8 = 64
9 X 8 = 72
10 X 8 = 80

What Goes Down, Must Come Up

Here's a saying that always stuck with me when remembering the 9 times table. Have you ever heard someone say, "What goes up must come down"? It's a common phrase that reminds us that nothing can rise forever and that everything eventually comes back into balance.

For example, when you first hear a catchy song, you'll be singing it nonstop and then a few years later think to yourself, "What was that song again?" Even when you throw a ball up in the air, it will go up and inevitably come back down.

Today, we're going to do things in reverse for the 9 times table from 9 x 1 to 9 x 10. Sometimes, what goes down must come up! Let's bring out the 9 times table chart and unveil a pattern you will never forget.

1 X 9 =
2 X 9 =
3 X 9 =
4 X 9 =
5 X 9 =
6 X 9 =
7 X 9 =
8 X 9 =
9 X 9 =
10 X 9 =

Steps

Ready to get counting? Let's start from the top and write 0 to 9 all the way down.

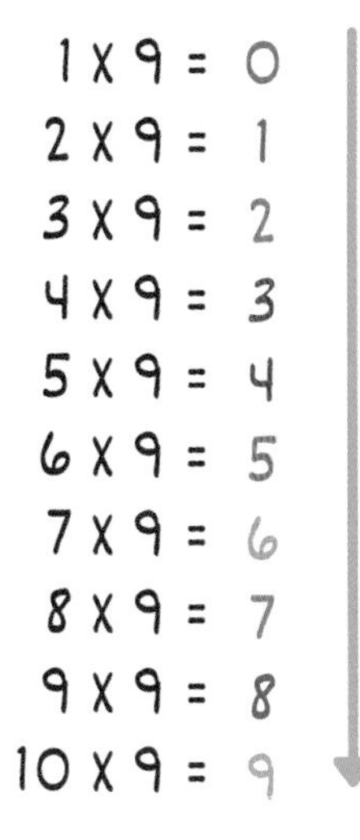

2

Let's keep the counting going! We're going to write 0 to 9 again, but this time going up. And just like that, you've completed your 9 times table!

1 X 9 = 09
2 X 9 = 18
3 X 9 = 27
4 X 9 = 36
5 X 9 = 45
6 X 9 = 54
7 X 9 = 63
8 X 9 = 72
9 X 9 = 81
10 X 9 = 90

Repeating Fractions Over 9

Have you ever noticed this pattern for proper fractions over 9? All of them will be a decimal that repeats the number above 9!

$$\frac{1}{9} = 0.11111111111111111...$$

$$\frac{2}{9} = 0.2222222222...$$

$$\frac{3}{9} = 0.3333333333...$$

$$\frac{4}{9} = 0.4444444444...$$

$$\frac{5}{9} = 0.5555555555...$$

$$\frac{6}{9} = 0.6666666666...$$

$$\frac{7}{9} = 0.777777777777...$$

$$\frac{8}{9} = 0.888888888888...$$

The 99 and 999 Sandwich

Once you know the 9 times table, you're well on your way to mastering the 99 and 999 times tables. The secret is simple—just use the same pattern you used for the 9 times table but sandwich a 9 or two in the middle. Let's whip out the 99 times table and give this a try!

1 X 99 =
2 X 99 =
3 X 99 =
4 X 99 =
5 X 99 =
6 X 99 =
7 X 99 =
8 X 99 =
9 X 99 =
10 X 99 =

Steps

Just like you did with the 9 times table, write 0 to 9 going from top to bottom.

1 X 99 = 0
2 X 99 = 1
3 X 99 = 2
4 X 99 = 3
5 X 99 = 4
6 X 99 = 5
7 X 99 = 6
8 X 99 = 7
9 X 99 = 8
10 X 99 = 9

② This time, write a line of 9s following your first row of numbers.

1 X 99 = 0 9
2 X 99 = 1 9
3 X 99 = 2 9
4 X 99 = 3 9
5 X 99 = 4 9
6 X 99 = 5 9
7 X 99 = 6 9
8 X 99 = 7 9
9 X 99 = 8 9
10 X 99 = 9 9

And finally, write 0 to 9 again, but this time going from the bottom to the top. And that's it! You've just created your own 99 times table in three simple steps.

1 X 99 = 0 9 9
2 X 99 = 1 9 8
3 X 99 = 2 9 7
4 X 99 = 3 9 6
5 X 99 = 4 9 5
6 X 99 = 5 9 4
7 X 99 = 6 9 3
8 X 99 = 7 9 2
9 X 99 = 8 9 1
10 X 99 = 9 9 0

Now that you know the secret pattern for the 99 times table, how would you create the 999 times table?

1 X 999 =
2 X 999 =
3 X 999 =
4 X 999 =
5 X 999 =
6 X 999 =
7 X 999 =
8 X 999 =
9 X 999 =
10 X 999 =

Steps

Write 0 to 9 going from top to bottom.

1 X 999 = 0
2 X 999 = 1
3 X 999 = 2
4 X 999 = 3
5 X 999 = 4
6 X 999 = 5
7 X 999 = 6
8 X 999 = 7
9 X 999 = 8
10 X 999 = 9

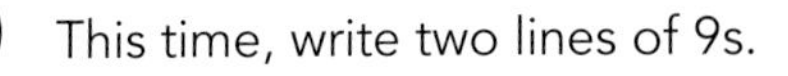

2. This time, write two lines of 9s.

1 X 999 = 0 9 9
2 X 999 = 1 9 9
3 X 999 = 2 9 9
4 X 999 = 3 9 9
5 X 999 = 4 9 9
6 X 999 = 5 9 9
7 X 999 = 6 9 9
8 X 999 = 7 9 9
9 X 999 = 8 9 9
10 X 999 = 9 9 9

Fill in 0 to 9 from the bottom to the top, and there you have the 999 times table!

1 X 999 = 0 9 9 9
2 X 999 = 1 9 9 8
3 X 999 = 2 9 9 7
4 X 999 = 3 9 9 6
5 X 999 = 4 9 9 5
6 X 999 = 5 9 9 4
7 X 999 = 6 9 9 3
8 X 999 = 7 9 9 2
9 X 999 = 8 9 9 1
10 X 999 = 9 9 9 0

No Charts? No Problem!

Are you ready to shake things up and try a new approach to multiplying 9 x 1 through 9 x 10? Give this a try!

Let's solve 9 x 7 using these two steps.

$$9 \times 7$$

Steps

1. First, let's solve for the tens digit. Here's the pattern—the tens digit in your answer will always be 1 less than the number you're multiplying 9 by.

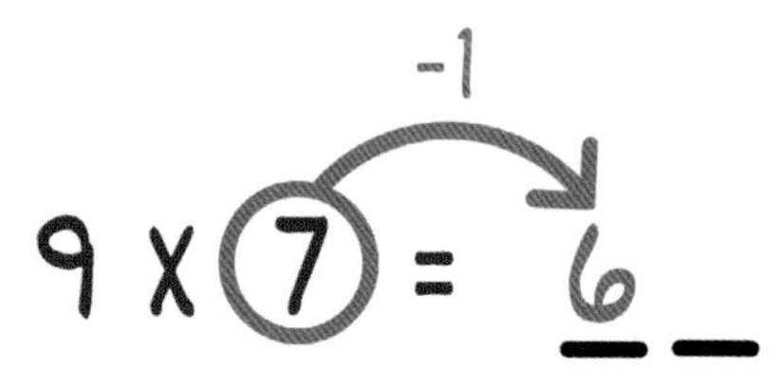

2. To solve for the ones digit, ask yourself, "What number added to the tens digit will equal 9?" In our example, the tens digit is 6 and adding 3 will equal 9. Therefore, 3 will be the ones digit and 63 will be your answer!

$$6 + \_ = 9$$

$$9 \times 7 = 63$$

Bonus

Let's try one more just so you get the hang of it!

$$9 \times 3$$

First, subtract 1 from 3 to get the answer's tens digit.

$$9 \times \textcircled{3} = \underline{2}\ \underline{\ \ }$$

What number added to the 2 you just wrote down will equal 9? It's 7! So the final answer for **9 x 3** will be 27.

$$2 + \_ = 9$$

$$9 \times 3 = \underline{2}\ \underline{7}$$

Multiply 9 Using Your Fingers

Do you like using your fingers to count? If you do, then you're going to love this next technique where we will multiply **9 x 1** to **9 x 10** using nothing but our fingers! All you need to do is open your hands in front of you with your palms facing up and label each of your fingers from 1 to 10.

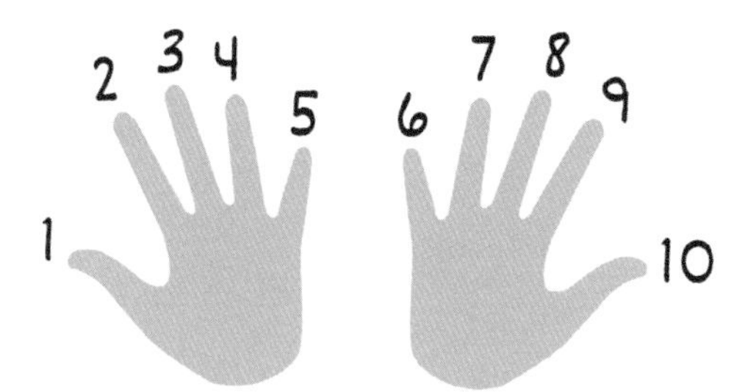

Looking good! You're all set to solve a problem. Let's start with **9 x 3**.

9 X 3

Steps

1. First ask yourself, "What number am I multiplying 9 by?" In this example, it's 3, so put down the finger labeled as 3. If you were multiplying **9 x 7** instead, you would put down the finger labeled as 7.

9 X 3 = __

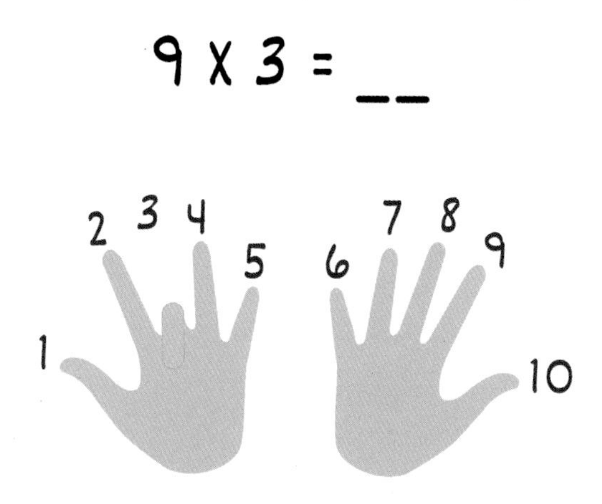

Next, count how many fingers are to the left of the finger you put down. This number will be your tens digit.

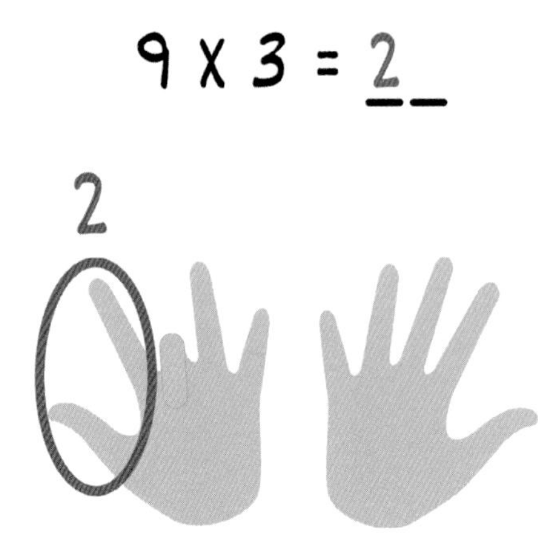

Then, count how many fingers are to the right of the finger you put down. This number will be your ones digit. Combining both the tens digit and the ones digit will yield your answer!

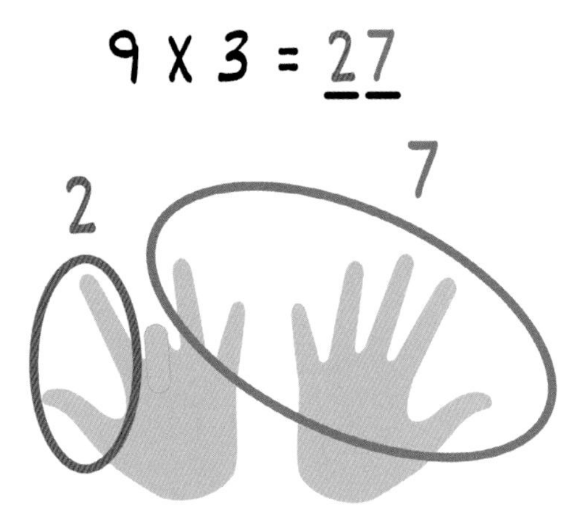

Works like magic, doesn't it? Let's try another example with **9 x 9** just so you get the gist of it. What's the first thing you need to do? Put down your 9th finger.

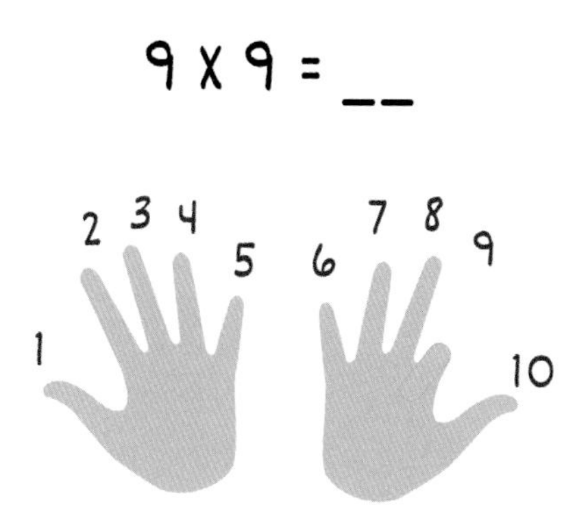

Count how many fingers are to the left of your 9th finger.

Count how many fingers are to the right of your 9th finger. Finally, combine the number of fingers on the left (8) and right (1) to get 81!

Multiply 6 to 10 Using Your Fingers

Let's take our finger math to the next level and multiply any number from 6 to 10 using our fingers. Open your two hands in front of you and label your fingers from 6 to 10 starting with 6 at your pinkies and 10 at your thumbs.

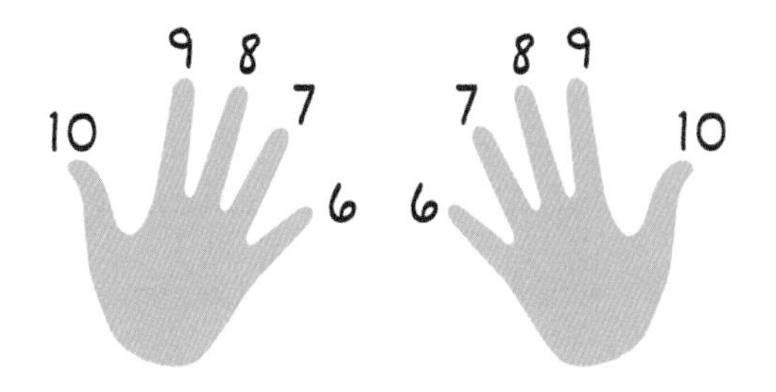

With your hands out and ready, let's try solving **7 x 8**.

7 X 8

Steps

First, turn your hands inward and touch the tip of your 7th finger on your left hand to the tip of the 8th finger on your right hand.

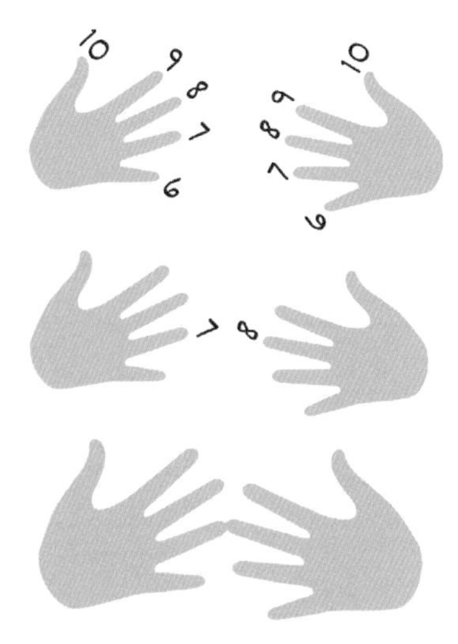

Next, count how many fingers are below the touching fingers and include the two touching fingers in your final count. Each finger represents 10, so if you count 5 fingers, that's 50.

5 fingers = 50

3 Then, count how many fingers are above your touching fingers, but this time don't include the fingers that are touching. In this example, there are 3 fingers on the left hand and 2 fingers on the right. Multiply these two numbers to get **3 x 2 = 6**.

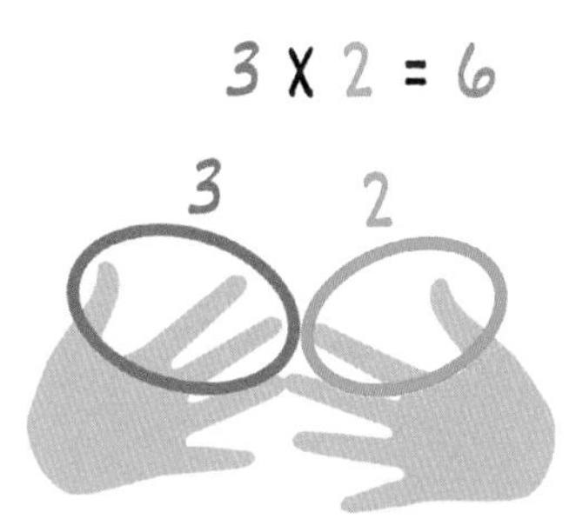

Finally, add 50 from the second step and 6 from the third step to get the answer of 56!

7 X 8 = 50 + 6 = 56

Bonus

Are you up for another fun round of finger multiplying? Let's try **6 x 7** and see how easy it can be. Touch the 6th finger on your left hand to the 7th finger on your right hand.

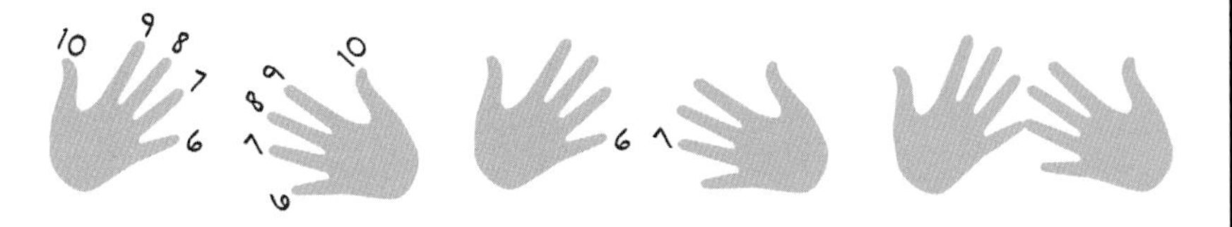

Next, count how many fingers are below the touching fingers including the two touching fingers. In this case, there are 3 fingers. Each finger represents 10, so 3 fingers is equal to 30.

3 fingers = 30

Now, count how many fingers are above the touching fingers, but this time don't include the touching fingers. In this example, there are 4 fingers on the left and 3 fingers on the right. Multiply these two numbers to get **4 x 3 = 12**.

4 X 3 = 12

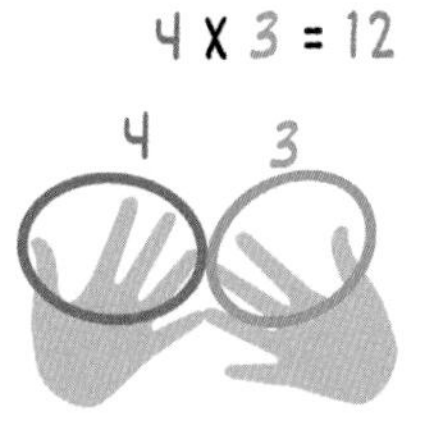

Finally, add 30 and 12 together to get the answer of 42! Easy, isn't it? Give it a try next time you multiply numbers from 6 to 10!

6 X 7 = 30 + 12 = 42

11 Comes in Pairs

The 11 times table is the easiest (and my favorite) times table. Why? Like socks, boots and earrings, 11s always come in pairs! Let's bring out the 11 times table chart and I'll show you what I mean.

1 X 11 =
2 X 11 =
3 X 11 =
4 X 11 =
5 X 11 =
6 X 11 =
7 X 11 =
8 X 11 =
9 X 11 =
10 X 11 =

Steps

For **11 x 1** to **11 x 9**, all your answers will have two digits, which will both be the number you are multiplying 11 by. Let's take **2 x 11** for example. Multiplying 11 by 2 is equal to 22!

1 X 11 = 11
2 X 11 = 22
3 X 11 = 33
4 X 11 = 44
5 X 11 = 55
6 X 11 = 66
7 X 11 = 77
8 X 11 = 88
9 X 11 = 99
10 X 11 =

2. For **11 x 10**, the first two digits will be the tens digit of 10 and the last digit will be the ones digit of 10. How about multiplying beyond **11 x 10**? In the Master the Art of Multiplication chapter (page 67), I'll show you a phenomenal trick you can use to multiply 11 by any two-digit or three-digit number, so stay tuned!

1 X 11 = 11
2 X 11 = 22
3 X 11 = 33
4 X 11 = 44
5 X 11 = 55
6 X 11 = 66
7 X 11 = 77
8 X 11 = 88
9 X 11 = 99
10 X 11 = 110

Simple Strategies for Addition and Subtraction

We all know the drill by now—big numbers are easier to work with once we break them down into smaller numbers (especially 0, 1, 2, 10 or 100) and combine them in a way that works best for us.

So, how would you add 735 and 213? Or subtract 253 from 378? There are plenty of ways you can break these big numbers down, but here are five ways I've used a lot as an engineer.

Place Value Chunks

This is the most common way to break down numbers when adding them—separate them into their place values (ones, tens, hundreds, thousands, etc.) and then add them up group by group.

$$735 + 213$$

$$= (700 + 30 + 5) + (200 + 10 + 3)$$

$$= (700 + 200) + (30 + 10) + (5 + 3)$$

$$= 900 + 40 + 8$$

$$= 948$$

But why is this technique so powerful? Let me show you using blocks. Here are two random piles of blocks. How many blocks are there if we combine the two piles?

It's going to take you a while to count the blocks in each pile. If we instead break each pile down into groups of ten blocks, we can easily see that there are 27 blocks in the first pile, 12 blocks in the second pile and 39 blocks in the third pile. This is why breaking your numbers down by their place value makes adding much easier!

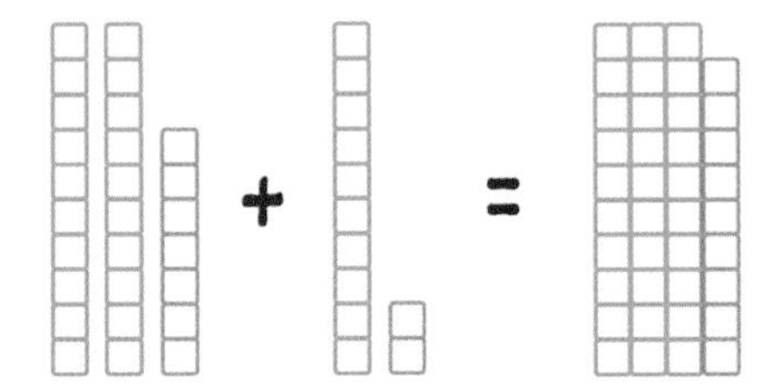

Shuffle the Deck

Another simple trick for adding big numbers is to group them into chunks that add up to nice round numbers. For example, if you see any two numbers whose last digits add to 10, try grouping them together first!

$$31 + 14 + 25 + 9 + 16$$

$$= (31 + 9) + (14 + 16) + 25$$

$$= 40 + 30 + 25$$

$$= 95$$

But your groups don't have to add up to 10. Every situation is different—get creative and look for patterns!

$$1 + 2 + 3 + 4 + 5 + 6 + 7 + 8 + 9 + 10$$

$$= (1 + 10) + (2 + 9) + (3 + 8) + (4 + 7) + (5 + 6)$$

$$= 11 + 11 + 11 + 11 + 11$$

$$= 55$$

Here's a fun challenge for you—can you figure out a way to quickly add up all the numbers from 1 to 100 by grouping them?

$$1 + 2 + 3 + 4 + \ldots + 97 + 98 + 99 + 100$$

$$= (1 + 100) + (2 + 99) + (3 + 98) + (4 + 97) + \ldots$$

$$= 101 + 101 + 101 + 101 + 101 + \ldots$$

$$= 50 \times 101$$

$$= 5050$$

And how did I know to multiply 101 by 50? Whenever you add consecutive numbers that start from 1 and end on an even number, the number of groups using the above pattern will always equal to the last even number divided by 2.

Give a Little, Take a Little

Here's a great way to add numbers that end in a 9. Convert them into nice round numbers that end in a zero by borrowing 1 from other numbers!

$$29 + 57$$

$$= (29 + 1) + (57 - 1)$$

$$= 30 + 56$$

$$= 86$$

The best part about this technique? It's not just limited to numbers that end in a 9. You can always borrow as much as you need from other numbers!

$$117 + 73$$

$$= (117 + 3) + (73 - 3)$$

$$= 120 + 70$$

$$= 190$$

Part I and Part II Subtraction

Let's move on to some subtraction techniques. Here is my go-to method for subtracting numbers less than 10. Try breaking down your subtraction problem into two parts. First, subtract your number down to a nice round number ending in zero. Then, subtract the rest!

32 - 7	173 - 9
32 - 2 = 30	173 - 3 = 170
30 - 5 = 25	170 - 6 = 164

Subtract Place Value Chunks

What if you're subtracting bigger numbers over 10? No worries, just break them down by their place value and then subtract them one at a time.

378 - 253

378 - 200 = 178

178 - 50 = 128

128 - 3 = 125

These are just a few methods for breaking down big numbers when adding and subtracting. As you work through the practice problems in this book, try experimenting with different techniques and see what works best for you. Get creative and have fun!

Add Odd Numbers in Seconds

$$1 + 3 + 5 + 7 + 9 + 11$$

Did you know that you can add $1 + 3 + 5 + 7 + 9 + 11$ without *actually* adding any of the numbers? When you are adding consecutive odd numbers starting from 1, a unique pattern begins to take form. Check this out!

Steps

Count the number of consecutive odd numbers that are added together. In our example, there are 6 odd numbers.

$$1 + 3 + 5 + 7 + 9 + 11$$

↑ ↑ ↑ ↑ ↑ ↑

1 2 3 4 5 6

Square that number to get the answer! $1 + 3 + 5 + 7 + 9 + 11 = 6^2 = 36$.

$$6^2 = 36$$

Let's try another example. What is $1 + 3 + 5 + 7 + 9 + 11 + 13 + 15$? This time, we are adding 8 consecutive odd numbers, so $1 + 3 + 5 + 7 + 9 + 11 + 13 + 15 = 8^2 = 64$.

Let's go one step further. How would you add $1 + 3 + 5 + \ldots + 97 + 99$? Instead of counting how many odd numbers you're adding, try doing this: Add 1 to the last number ($99 + 1 = 100$) and then divide it by 2 ($100 \div 2 = 50$). This means there are 50 odd numbers between 1 and 99 inclusive and so $1 + 3 + 5 + \ldots + 97 + 99 = 50^2 = 2{,}500$.

Practice

$1 + 3 + 5 + 7 + 9 + 11 + 13 + 15 + 17 + 19 + 21$

Sum of odd numbers from 1 to 199

Sum of all odd numbers up to 2,007

Why Does This Work?

Let me introduce you to the concept of *arithmetic progression*. It's a bit lengthy and involves a fair amount of algebra, but if you're up for a challenge, then go for it! So what exactly is an arithmetic progression? It's basically a series of numbers that increase by the same amount each time. For example, 5, 8, 11, 14, 17 is an arithmetic progression because each number is 3 more than the previous one. However, 1, 2, 7, 98 is not an arithmetic progression because the increase from 2 to 7 is different than the increase from 7 to 98.

So how is this related to adding consecutive odd numbers? Well, it turns out that adding consecutive odd numbers (you know, 1, 3, 5, 7, 9 . . .) is an arithmetic progression that increases by 2 every time. And if you would like to add up all the numbers in an arithmetic progression, there's a formula for that: $S_n = (n/2)(2a + (n - 1)d)$. Here "$S_n$" is the sum, "n" is how many numbers you're adding, "a" is the first number and "d" is the difference between each number.

$$S_n = \frac{n}{2}[2a + (n - 1)d]$$

S_n = Sum of arithmetic progression

n = number of terms

a = the first term

d = the difference between each number

When we add consecutive odd numbers, we can set "a" as 1 and "d" as 2. Then, if we plug those values into the formula from earlier, it simplifies down to $\mathbf{S_n = n^2}$! This means that the sum of "n" number of consecutive odd numbers will be n^2.

Feel like skipping the algebra today? There's another way to prove this that's way more fun—with shapes! Let's grab some blocks and arrange them in an L shape to create a few odd numbers.

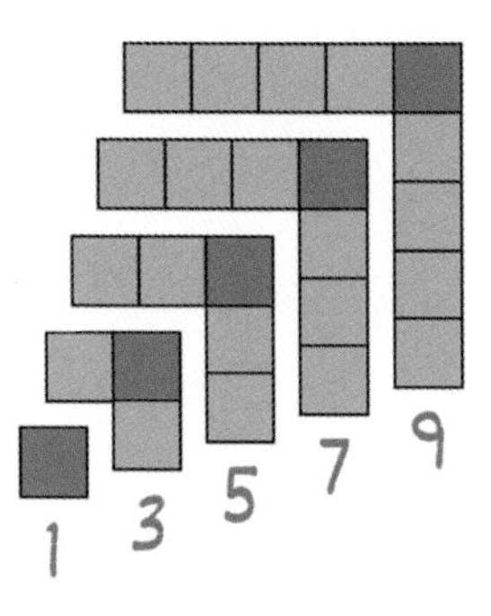

By simply connecting L-shaped blocks together, guess what you'll create? A square! The number of blue blocks in your square equals the number of odd numbers you are adding together. That's the secret to understanding the sum of consecutive odd numbers: It's just the square of the number of consecutive odd numbers that are being added together.

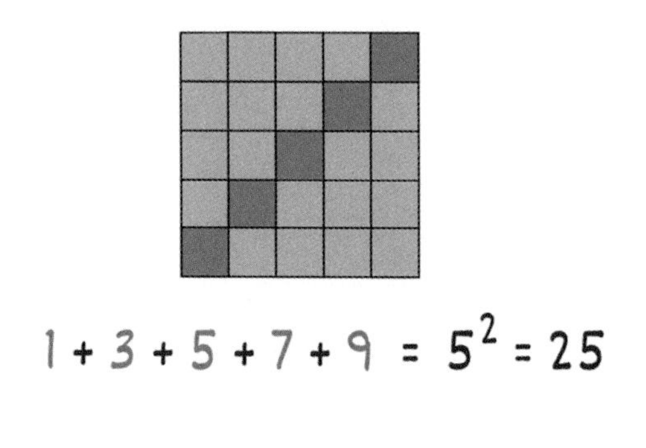

Add Even Numbers Mentally

You've learned the trick to instantly add consecutive odd numbers like $1 + 3 + 5 + 7 + 9$, but what about even numbers? Observe the magic that happens when we add $2 + 4 + 6 + 8 + 10 + 12$!

$$2 + 4 + 6 + 8 + 10 + 12$$

Steps

Count the number of even numbers being added.

$$2 + 4 + 6 + 8 + 10 + 12$$

↑ ↑ ↑ ↑ ↑ ↑

1 2 3 4 5 6

2. Multiply that number by 1 more than itself and that'll be your answer! In this example, we have 6 even numbers. From there, 1 more than 6 is 7 and when multiplying 6×7 we get 42! Therefore, $2 + 4 + 6 + 8 + 10 + 12 = 42$.

$$6 \times 7 = 42$$

↑

6 + 1

Let's try something a little harder. How would you add all consecutive even numbers up to 200 ($2 + 4 + 6 + \ldots + 196 + 198 + 200$)?

Instead of individually counting how many even numbers you see, remember the trick we learned on page 55. Divide 200 by 2 ($200 \div 2 = 100$) to get 100 even numbers between 2 and 100. Now you can multiply 100 by one more than itself to get the answer ($100 \times 101 = 10{,}100$).

Practice

2 + 4 + 6 + 8 + 10 + 12 + 14 + 16 + 18

Sum of even numbers from 2 to 20

Sum of all even numbers up to 1,000

Why Does This Work?

Remember the arithmetic progression formula from earlier when we added consecutive odd numbers?

$$S_n = \frac{n}{2}[2a + (n - 1)d]$$

S_n = sum of arithmetic progression
n = number of terms
a = the first term
d = the difference between each number

Well, adding consecutive even numbers (like 2, 4, 6, 8, 10 . . .) is also an arithmetic progression that increases by 2 every time, but this time we're starting at 2 instead of 1. This means that "a" will be 2 ("d" remains as 2). If we plug those values into the formula, it'll simplify down to $S_n = n(n + 1)$. This means that the sum of "n" consecutive even numbers is $n(n + 1)$!

Subtract Big Numbers without Borrowing

$$\begin{array}{r} 8000 \\ -\ 3729 \\ \hline \end{array}$$

Here's a creative way for subtracting from a large number that ends in zeros, like 500, 82,000 or 108,000. It's easy and fast, but it won't work as well if the zeros are not at the end, like 108 or 900,582. And the cherry on top? You won't need to borrow anymore!

Steps

1. Start by subtracting 1 from both of your numbers. For example, 8,000 will become 7,999 and 3,729 will become 3,728.

$$\begin{array}{r} 8000-1 \\ -\ 3729-1 \\ \hline \end{array} \longrightarrow \begin{array}{r} 7999 \\ -\ 3728 \\ \hline \end{array}$$

2. Now just subtract the digits in each place value as you normally would going from right to left. Start by subtracting **9 - 8** in the ones place and work your way to the thousands place. When you're done, you'll have your answer!

$$\begin{array}{r} 7999 \\ -\ 3728 \\ \hline 1 \end{array} \longrightarrow \begin{array}{r} 7999 \\ -\ 3728 \\ \hline 4271 \end{array}$$

Practice

$$\begin{array}{r} 700 \\ -\ 83 \\ \hline \end{array}$$

$17{,}000 - 936$

$-238 + 5000$

Why Does This Work?

When you subtract numbers, all you're doing is finding the distance between them on a number line. For example, **8 - 2 = 6** because there are 6 units between 2 and 8.

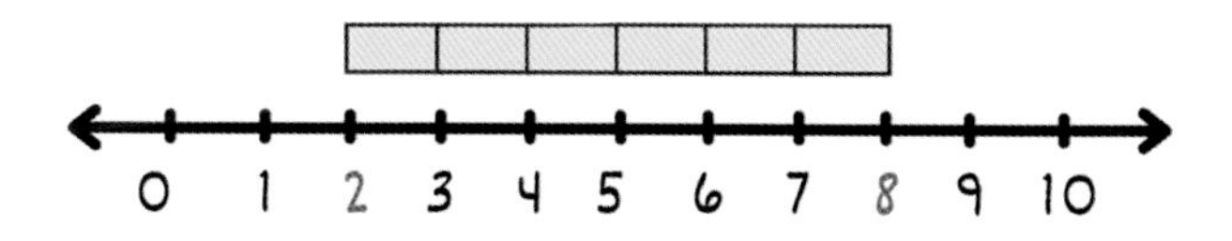

Now if we subtract 1 from both 8 and 2, then the expression **8 - 2** becomes **7 - 1**. However, the answer will remain as 6 because the distance between 7 and 1 is the same as between 8 and 2. All we did was shift everything down by 1!

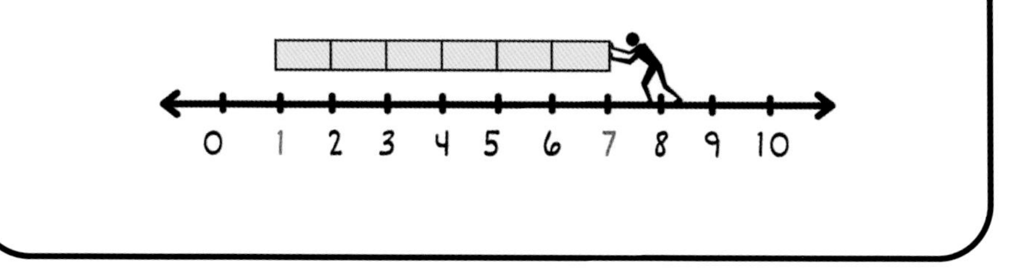

How Much Do You Make After Taxes?

Let's say you earn a salary of $50,000 one year (nice!). Unfortunately, you don't get to keep the full $50,000 because you need to deduct your taxes and health insurance from your employer. If your taxes and health insurance add up to $16,724, how much money do you take home at the end of the year?

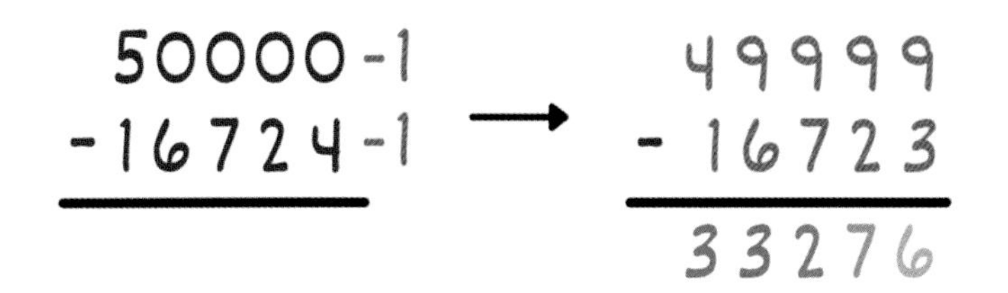

You get to keep a grand total of $33,276 in your pocket! But does that feel a little low? That's the reality of taxes, my friend, but not to worry! You can always find ways to make the most of your hard-earned cash.

Subtract by Adding?!

How would you solve a simple subtraction problem like **20 - 11** in your head? The most straightforward method is to start at 20 and subtract 11 to get 9. But did you know there's another way to solve this? You could also start at 11 and count up 9 to reach 20.

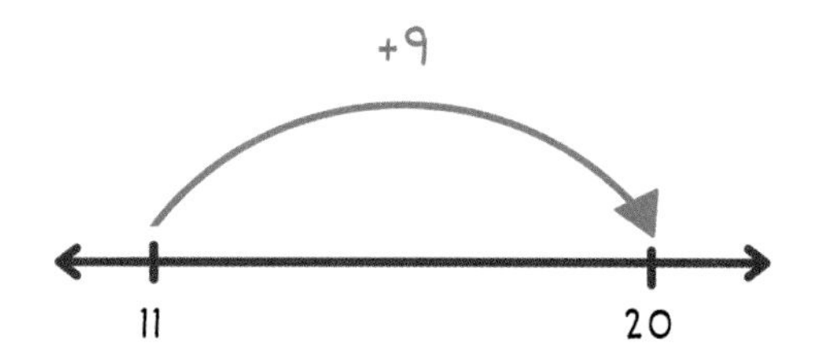

Subtraction is all about the distance between two numbers. It's simple for us to calculate the distance between 11 and 20 when subtracting **20 - 11**, but what about larger numbers such as **94 - 37** or **913 - 228**? Don't worry, I've got a handy trick for you!

94 - 37

Steps

1. Let's solve **94 - 37** by counting up. First start at the lower number, 37, and count up to the nearest number that ends in a zero, which is 40.

Now take 40 and start a new row. We're going to count up again but in groups of 10 until we reach the closest number below 94 that ends in a zero, which is 90.

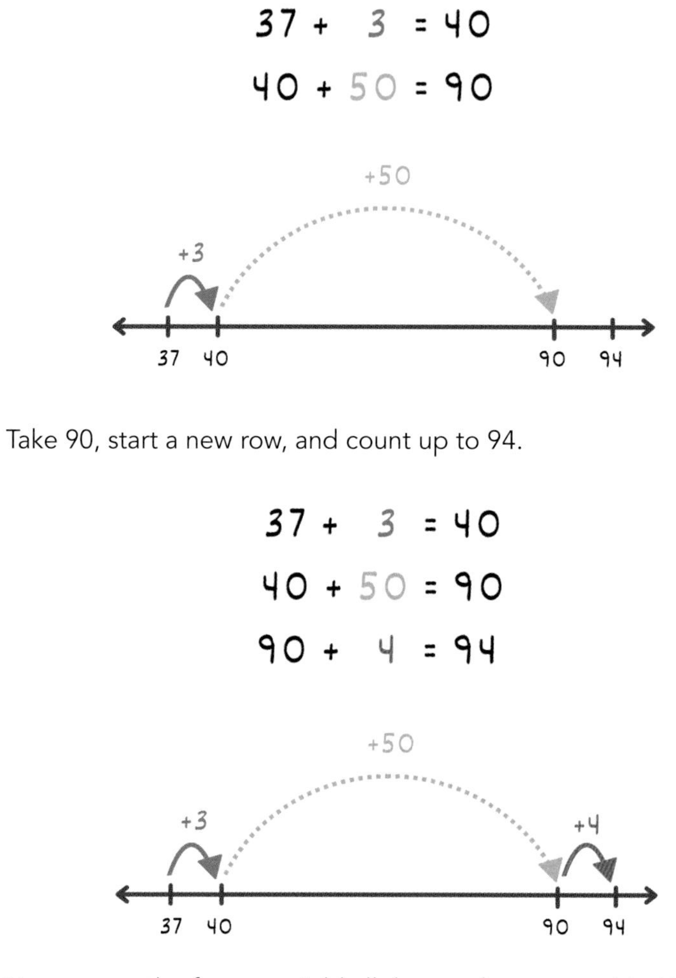

③ Take 90, start a new row, and count up to 94.

④ Now comes the fun part. Add all the numbers you added in each row to get the answer for **94 - 37**!

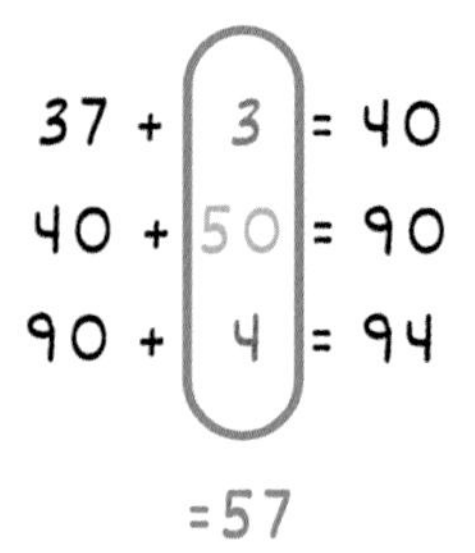

Practice

52 - 17

1,234 - 321

3,920 - 1,242

Why Does This Work?

Subtracting **94 - 37** in one step is a challenge for our brains to process. By breaking the problem down into smaller subtraction problems (**40 - 37**, **90 - 40** and **94 - 90**), the problem becomes much more manageable. That's because adding groups of 10, like from 40 to 90, is much faster than adding every number individually from 37 to 94.

Think of it like this—you would never walk from New York City to Boston; it would take you three days! It will be much faster to take the train most of the way and only walk a short distance to and from the train stations.

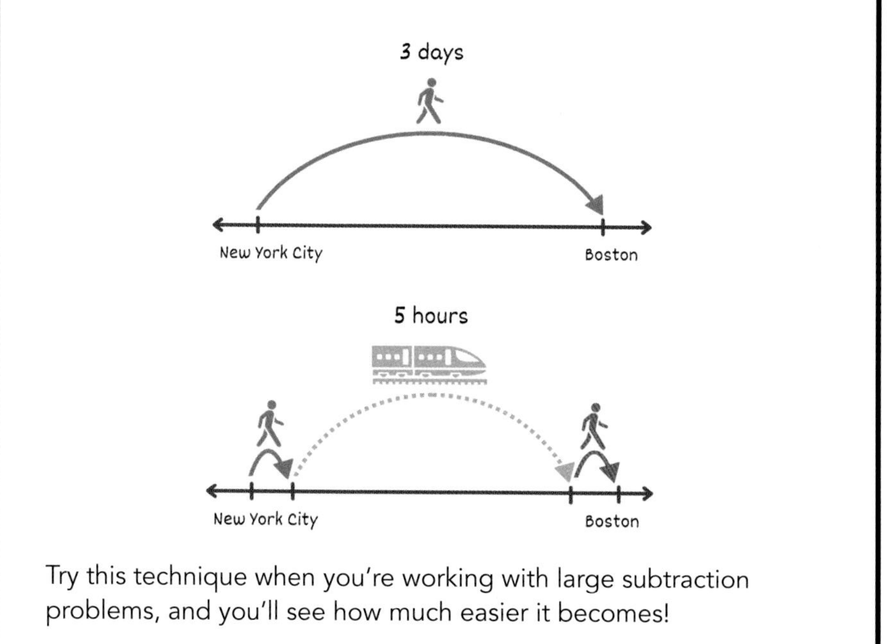

Try this technique when you're working with large subtraction problems, and you'll see how much easier it becomes!

How Many Miles Do You Have Left on a Road Trip?

When I started my job at NASA in Cape Canaveral, I drove my car all the way down from New York to Florida. The whole trip was 913 miles (1,469 km), so I took rest stops in Washington, DC, and Charleston, South Carolina. If I drove 228 miles (367 km) from New York to Washington, DC, how many miles did I have left until Cape Canaveral?

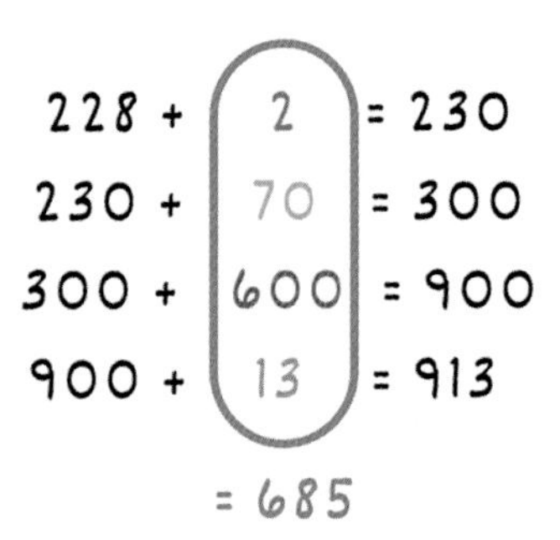

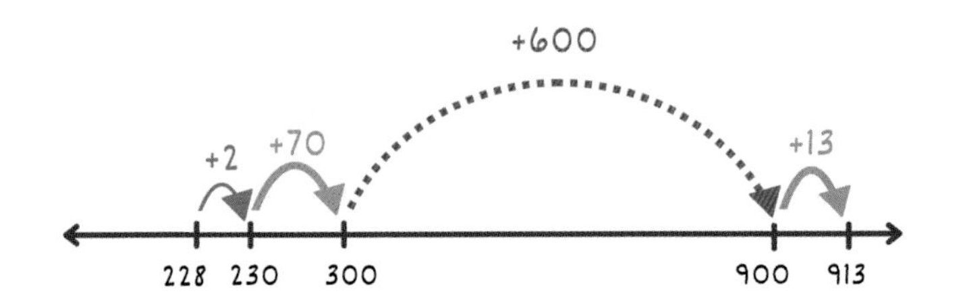

At Washington, DC, I still had 685 miles (1,102 km) to go until Cape Canaveral!

Master the Art of Multiplication

Can you quickly solve these two multiplication problems in your head?

$$9 \times 4 \qquad 14 \times 4$$

Because you have your 9 times table memorized, $9 \times 4 = 36$ is a piece of cake. But how about 14×4? How would you quickly solve that? You don't need to have the 14 times table memorized; simply remember this concept:

Multiplication is just repeated addition.

Once you've wrapped your head around this, you can easily tweak any multiplication problem to become easier. For example, if I spend $4 on coffee every day, how much would I spend on coffee over a total of 2 weeks? To solve this, you'll need to multiply 14×4.

$$14 \times 4$$

14 times of 4

$$4 + 4 + 4 + 4 + 4 + 4 + 4 + 4 + 4 + 4 + 4 + 4 + 4 + 4$$

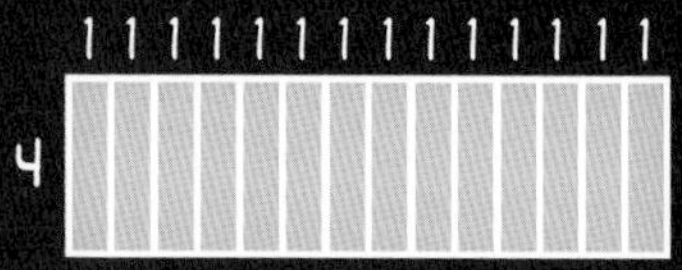

1 1 1 1 1 1 1 1 1 1 1 1 1 1

4

Instead of adding 4 fourteen times, you can group some 4s together and add the groups. For example, you can group ten 4s together to get 40 and four 4s together to get 16 (continued on the next page).

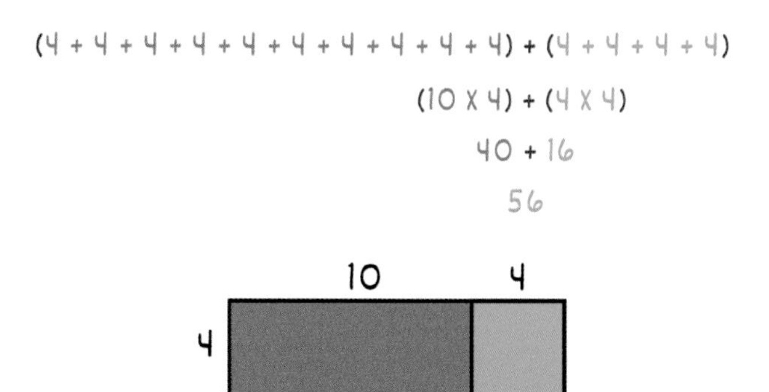

What are other ways you can break down 14 x 4 and solve it in your head?

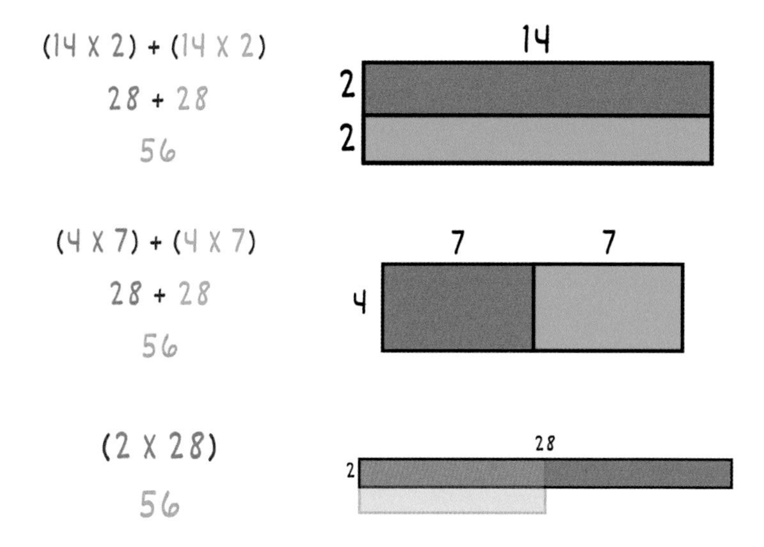

The party doesn't stop here—there are still other ways you can solve 14 x 4. In this section, you'll discover ten mind-blowing ways to break down those tricky multiplication problems and solve them in seconds. The most important thing is to have fun and unleash your inner creativity. Remember, a problem may have only one answer, but there can be many paths to get there!

Pro Tip: The word "of" means to multiply. For example, 20% of 50 means 20% x 50*, just like how two-thirds of 30 means* $\frac{2}{3}$ x 30*. If you've spent one-fourth of a $100 gift card, you've spent* $\frac{1}{4}$ x $100*, which is $25.*

Multiplying by 5? Do This!

Let's kick off our multiplication tricks with the one I think you'll use the most. How would you multiply **18 x 5** in your head? Here's how to solve this faster than you can say "**18 x 5**"!

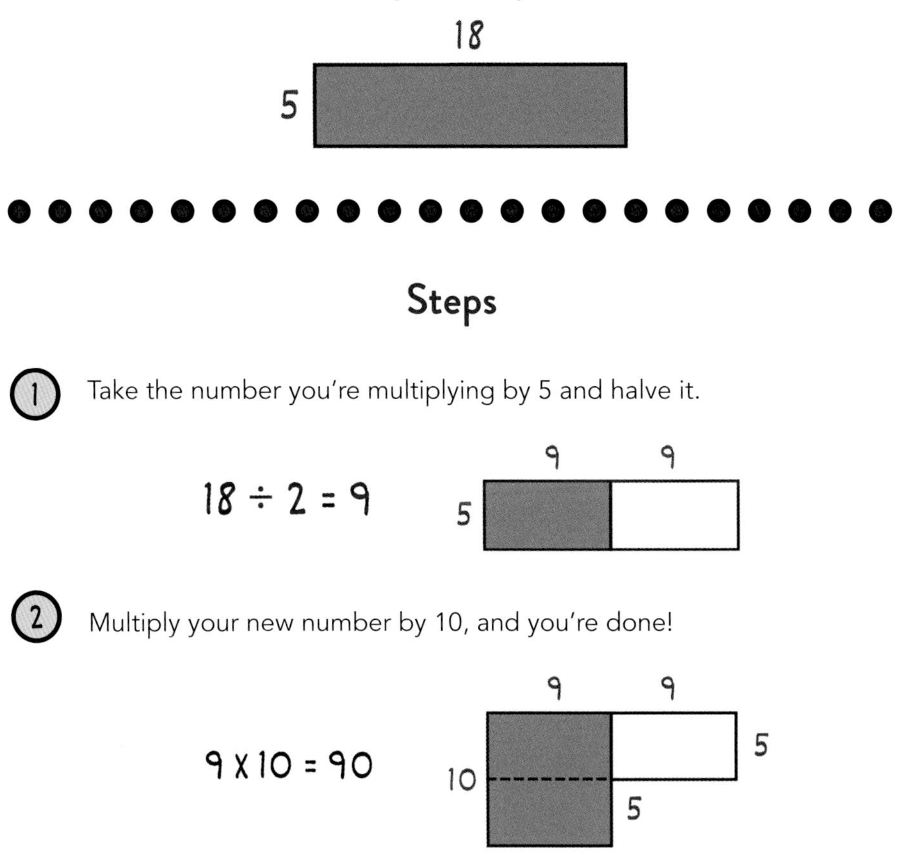

Steps

1. Take the number you're multiplying by 5 and halve it.

$18 \div 2 = 9$

2. Multiply your new number by 10, and you're done!

$9 \times 10 = 90$

You can use this technique for any number multiplied by 5. Trust me, you'll be using it all the time in the future, it's that useful!

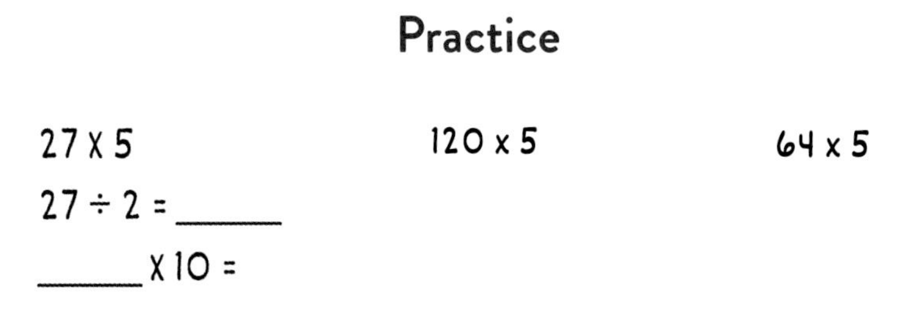

Practice

27 x 5

27 ÷ 2 = ______

______ x 10 =

120 x 5

64 x 5

Why Does This Work?

Remember, the easiest numbers for your brain to work with are 0, 1, 2, 10 and 100. Because **$5 = 10 \div 2$**, multiplying a number by 5 is the same as multiplying it by 10 and dividing it by 2. It will always be easier to solve two steps using 0, 1, 2, 10 and 100 than one step using difficult numbers.

How Many Days Do You Work in a Year? Let's Find Out!

Let's put this trick to the test! Let's say you work every weekday for 48 weeks a year. To calculate the number of days you work a year, first divide 48 by 2 (**$48 \div 2 = 24$**) and then multiply 24 by 10 (**$24 \times 10 = 240$**). You work a total of 240 days each year!

This 11 Trick Will Blow Your Mind

Can you quickly recall **3 x 11** and **8 x 11**? I'm sure you can! But how about bigger numbers like **829 x 11** or **2,357 x 11**? If you're tempted to reach for a calculator, here's a trick to multiply them without even lifting a finger!

72 X 11

Steps

1. Before we begin, here's an important thing to keep in mind—whenever you multiply 11 by a two-digit number, you can split your answer into three parts and solve for each. Now the first thing you need to do is to copy the first digit of the number multiplied by 11 into the answer. In our example, the first digit in 72 is 7, so copy that over!

   72 X 11 = 7 _ _

2. Next, copy the second digit over to your answer's last digit. In this case, the second digit in 72 is 2.

   72 X 11 = 7 _ 2

3. Finally, add together the first and last digits in your answer to get its middle digit!

   72 X 11 = 792

   7+2 = 9

This trick works for any two-digit number, but if the sum of your middle digit is above 9, you'll need to carry over the tens place. Here's what I mean—take a look at **85 x 11**. We first copied the first digit (8), then copied the last digit (5) and finally added them together to get the middle digit (**8 + 5 = 13**). However, because 13 is greater than 9, you'll need to carry the tens digit (1) over to the answer's hundreds place. This will bump 8 up to 9 and your new answer will be 935 (not 835 or 8,135)!

$$85 \times 11 = \overset{+1}{8}\,3\,5 = 935$$

$$8 + 5 = 13$$

What happens when you multiply 11 by a three-digit number? You can use same technique we just did!

527 X 11

Steps

1. When multiplying 11 by a three-digit number, you can split your answer into four parts and solve for each. Just like before, copy the first digit of the number multiplied by 11 into the answer.

$$527 \times 11 = 5\_\_\_$$

2. Next, copy over the last digit.

$$527 \times 11 = 5\_\_7$$

3. Now for the answer's middle digits! To solve for the second digit, simply add the first and second digit in your original number.

$$527 \times 11 = 57\_7$$

$$5 + 2 = 7$$

To get the answer's third digit, add the second and third digit in your original number.

527 X 11 = 5797

2+7 = 9

Cool, isn't it! You may use this technique to multiply 11 by a four-digit, five-digit or even a twenty-digit number. All you need to do is copy the first digit, copy the last digit and then keep adding together every consecutive digit to get the answer's middle digits!

Practice

53 X 11 = _ _ _ 86 X 11 = _ _ _ 7253 X 11 = _ _ _ _ _

Why Does This Work?

The reasoning behind this trick is simple. First, would you agree we can break 11 down into **10 + 1**? Now, if we multiply **23 x 11**, we can break the problem down into **(23 x 10) + (23 x 1)**, which simplifies down to **230 + 23**. From this we can see that the first digit will remain as 2, the last digit as 3, and the middle digit will be **2 + 3**!

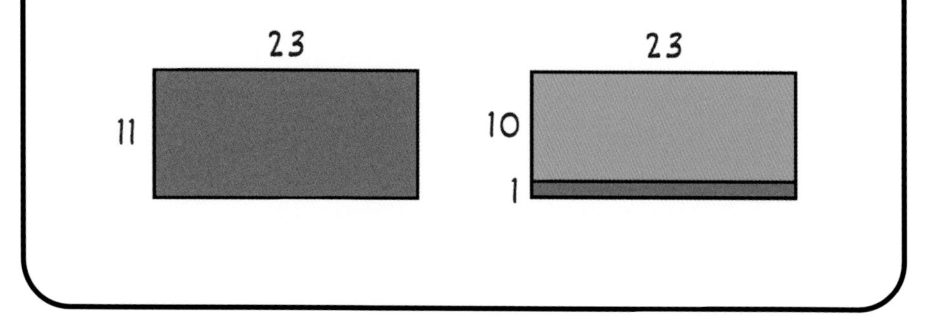

Hotel Budgeting on Your Dream Trip

You've been dreaming of taking a long vacation in Italy during the summertime and now it's finally happening! Your trip will last 11 days and the average hotel cost is $154 per night. How much should you budget for hotel costs for the entire trip? To calculate this, you simply need to multiply the average rate of $154 by the number of nights (**154 x 11**).

Using the trick in this chapter, you can solve this calculation in four easy steps! First, copy the first digit in 154.

$$154 \times 11 = 1\_\ \_\ \_$$

Then, copy the last digit in 154.

$$154 \times 11 = 1\_\ \_\ 4$$

For the answer's second digit, add the first digit (1) and the second digit (5) in 154.

$$154 \times 11 = 16\_\ 4$$

$$1+5=6$$

And for the answer's third digit, add the second digit (5) and the third digit (4) in 154.

$$154 \times 11 = 1694$$

$$5+4=9$$

And there we have it! The total cost of hotel fees for your 11-day vacation in Italy will be $1,694. It's time to begin saving up for this exciting trip!

11 x 11? How about 11111111 x 11111111?

While we're talking about the number 11, let me show you a fascinating pattern before moving on to the next multiplication trick. Have you seen what happens when you multiply numbers whose digits are all 1s? The answer will start with a 1 end with a 1, but the real magic happens in between!

The answer's digits will go in order from 1 up to a particular number and then back down to 1. And what significance do you think this particular number has? It's the number of digits in the two numbers you're multiplying together. There is something so visually satisfying about this perfectly symmetrical pyramid of numbers, don't you agree?

1 X 1 = 1

11 X 11 = 121

111 X 111 = 12321

1111 X 1111 = 1234321

11111 X 11111 = 123454321

111111 X 111111 = 12345654321

1111111 X 1111111 = 1234567654321

11111111 X 11111111 = 123456787654321

111111111 X 111111111 = 12345678987654321

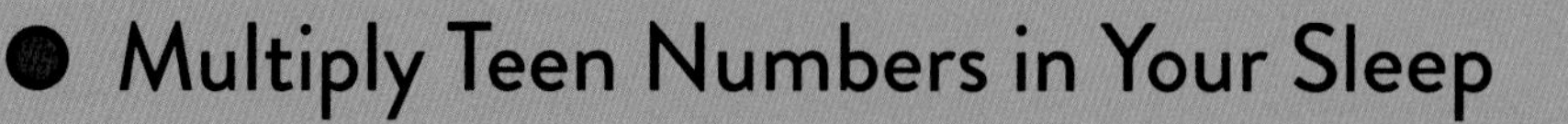

Multiply Teen Numbers in Your Sleep

Do you enjoy jamming out to music while you work? I always plug my earphones in and play songs on Spotify to keep myself motivated and focused. But this doesn't come cheap! I pay a Spotify subscription of \$12.99 a month (but let's keep it simple and round it up to \$13). The big question is, how much do I pay each year?

How would you calculate **\$13 x 12** in your head? I have a little trick that makes it super easy to multiply any two numbers between 11 and 19. Ready to give it a go?

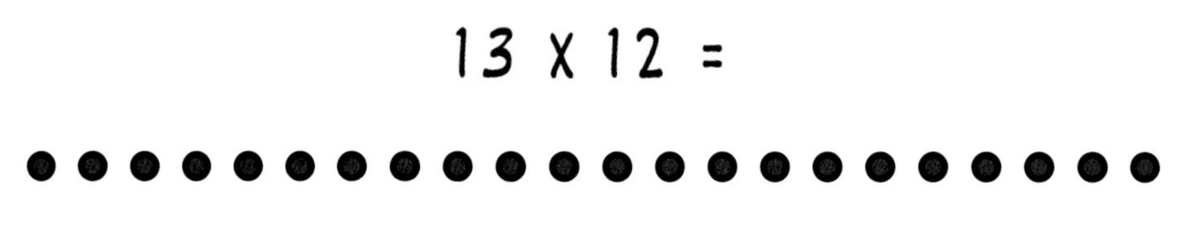

$$13 \times 12 =$$

Steps

Whenever you multiply two numbers between 11 and 19, your answer will always have three digits.

$$13 \times 12 = \_\_\_$$

2. In this step, we will solve for the first two digits (hundreds and tens place) in the answer. First, pick one of your numbers, grab the ones digit and add it to the other number. For example, for **13 x 12** you can either take the 2 from 12 and add it to 13 or take the 3 from 13 and add it to 12. Either way, these will add to 15, which will be the first two digits of your answer.

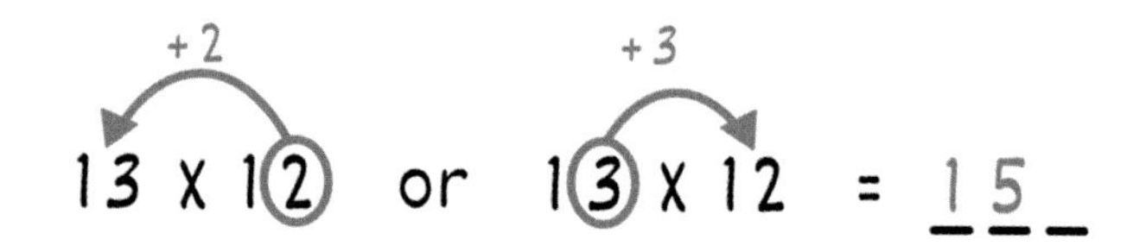

3. Next, let's solve for the third digit in your answer. To do this, grab the ones digit in both of your numbers and multiply them. This means I spend **$13 x 12 = $156** a year on my music subscription!

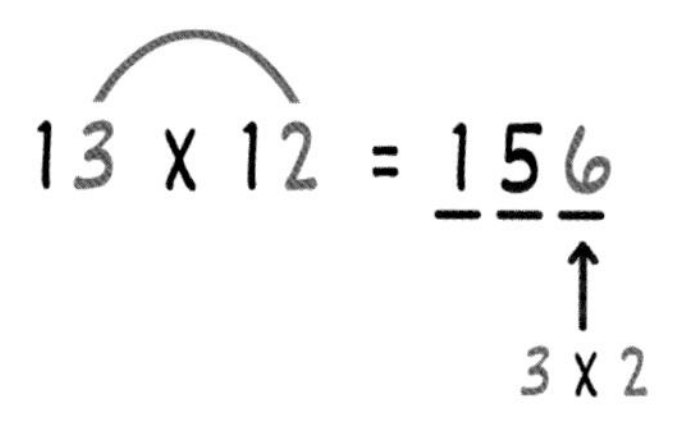

Easy, isn't it? However, if your answer's third digit adds up to 10 or above, you'll need one extra step involving a bit of carrying. Let's try this out in the next example.

12 X 16 = ___

Steps

1. Just like before, let's first solve for the first two digits. Pick one of your numbers, grab the ones digit and add it to the other number.

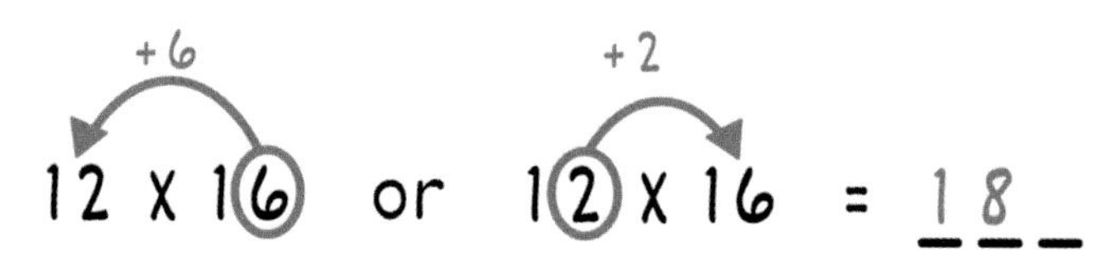

2. Next, let's solve for the third digit in the answer by grabbing the ones digits in both numbers and multiplying them. Because **6 x 2 = 12** and 12 is above 9, you'll need to carry the 1 in 12 over to the answer's tens place.

$$16 \times 12 = \underline{1}\,\underline{8}\,\underline{2}$$

+1 (carried to the tens place); $6 \times 2 = 12$

3. Add the 1 you just carried over to the number in that place value to get the final answer of 192!

$$16 \times 12 = \underline{1}\,\underline{9}\,\underline{2}$$

Practice

14 x 15

13 x 18

17 x 19

Why Does This Work?

Let's look at our first example, **13 x 12**. Before moving further, let's break our numbers down into their place values (**13 = 10 + 3** and **12 = 10 + 2**). To solve **13 x 12**, we can multiply their place values together and add them: **13 x 12 = (10 + 3)(10 + 2) = (10 x 10) + (10 x 3) + (10 x 2) + (3 x 2)**.

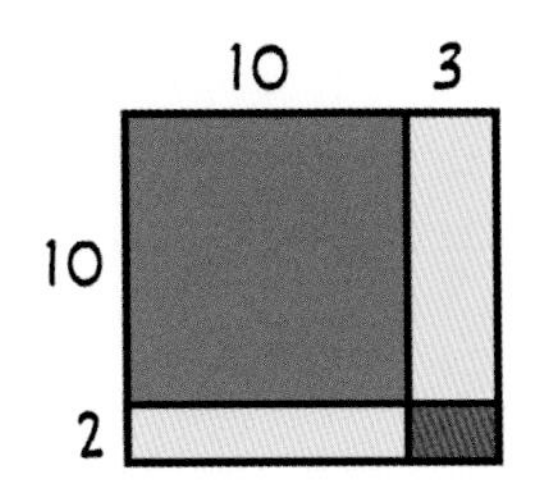

How does this relate to the trick? In the first step, we grabbed the ones digit from one number and added it to the other number. All we did here was multiply all the tens digits and the ones digits in 13 and 12 together: **(10 x 10) + (10 x 3) + (10 x 2)**.

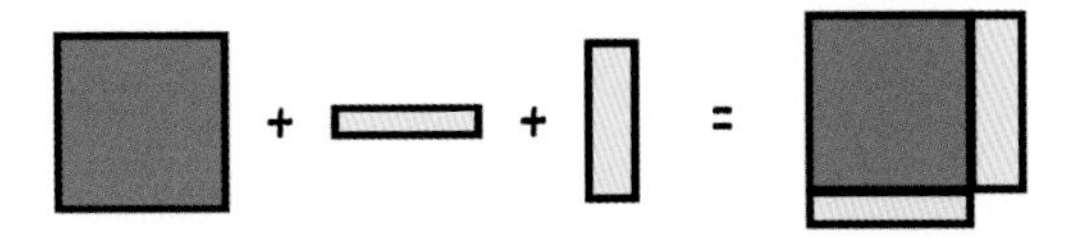

In the second step, we added the final piece to the puzzle and multiplied the ones digits together **(3 x 2)**.

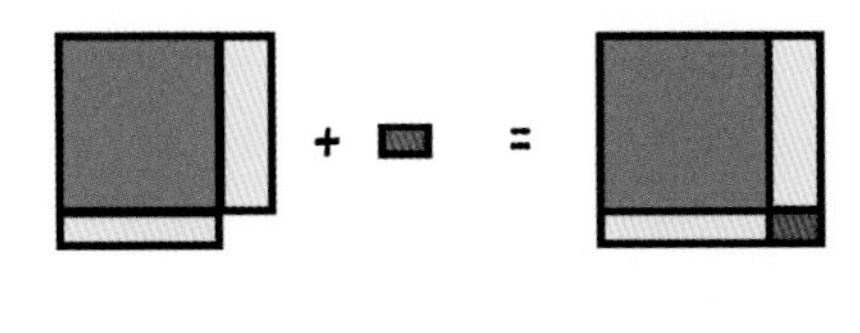

Two Apart? Too Easy!

When you multiply two numbers that are only two units apart, something truly unique and exciting happens. If you're familiar with the *difference of squares* concept, you'll already have a sneak peek of what's to come. If not, you're in for a treat!

$$101 \times 99$$

Steps

Find the number in between the two numbers that you're multiplying. In our example, 100 is between 101 and 99.

99...100...101

Then, take that number and square it!

$$100^2 = 10{,}000$$

Finally, subtract 1 to get your answer.

$$10{,}000 - 1 = 9{,}999$$

Cool, isn't it? Now, this trick works on any two numbers that are two units apart, but it does have its limitations. Let's take **77 x 79** as an example. Using this trick, you'd find the number in between them (78), square it (78^2) and subtract 1 (78^2 - 1). But do you know what 78^2 is off the top of your head? I know I don't! That's why this trick works best when your two numbers are around a number you can easily square, like multiples of 10 (20, 30, 40, 100, 200, 300, etc.). That way, you can easily visualize the answer and save yourself some mental gymnastics!

Practice

13 x 11

79 x 81

301 x 299

Why Does This Work?

Let's multiply two easy numbers like **9 x 7**. Would you agree that 9 is just (**8 + 1**) and 7 is just (**8 - 1**)? Therefore, **9 x 7** is equal to **(8 + 1)(8 - 1)**. Now I'm going to prove to you how **$(8 + 1)(8 - 1) = 8^2 - 1^2 = 8^2 - 1$** with a bit of algebra and geometry. Let's represent 8 as "a" and 1 as "b" to prove that **$(a + b)(a - b) = (a^2 - b^2)$**. First, create a square whose sides have a length of "a." The area of this square is a^2.

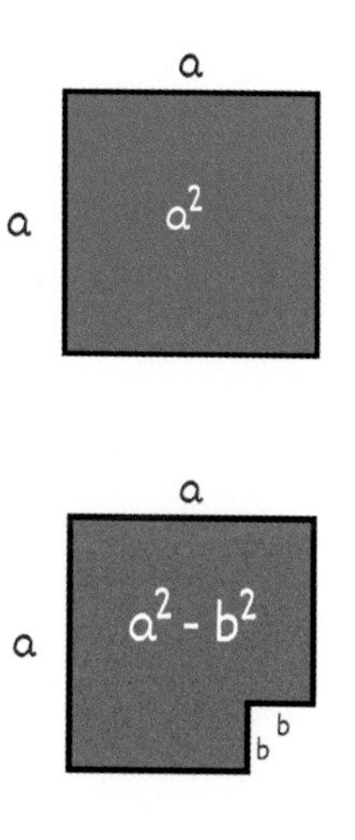

To get $(a^2 - b^2)$, let's cut out a small square (b^2) from our big square (a^2).

Now this square is looking a little sad. How could you slice it up and move pieces around to make it a nice-looking rectangle again? Well, you could shave off a part of the bottom and stick it on the side. And what is the area of this new rectangle? It's **$(a + b)(a - b)$**!

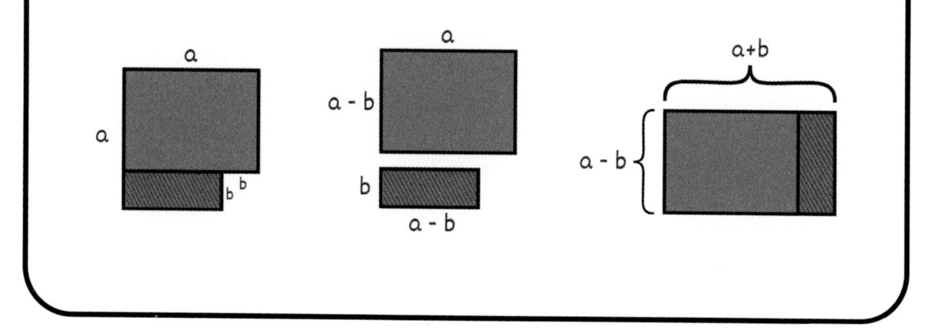

Did You Get the Right Size Rug?

Let's say you have a snug squarish closet space with an area of around 3,000 in^2 (19,354.8 cm^2). You're doing a little online shopping and fall in love with a rug that is 49 by 51 inches (124.5 by 129.5 cm) with frills on both ends. Doing a quick calculation, will this rug fit in your closet?

To solve this, let's multiply 49 by 51 inches.

$$49 \times 51 = 50^2 - 1 = 2,499$$

This means that your rug is 2,499 in^2 (16,122.75 cm^2) and will fit nicely in your closet!

The 7-11-13 Coincidence

Can you guess what happens when you take any three-digit number and multiply it by 7, 11 and 13? I call this the 7-11-13 coincidence, but is it just a coincidence? Let's find out.

Steps

1. Pick any three-digit number. In this example, let's pick 238.

238

2. Take your number and multiply it by 7, then 11 and then 13. And just like magic, your number will turn into a six-digit number with its first three digits repeated.

238 X 7 X 11 X 13 = 238238

Cool, isn't it? Let's see more examples of this.

956 X 7 X 11 X 13 = 956956

123 X 7 X 11 X 13 = 123123

700 X 7 X 11 X 13 = 700700

Why Does This Work?

Although this pattern may seem random at first, there's a clear explanation behind it. Just think about it for a moment—what happens when you multiply **7 x 11 x 13**? It equals 1,001! And here's the kicker: Whenever you multiply a three-digit number by 1,001, its digits repeat in the format **abc x 1,001 = abc,abc**.

But why does this happen? Let's break down 1,001 by its place values (**1,001 = 1000 + 1**). When you multiply a number like 238 by 1,001, it's the same as multiplying **(238 x 1,000) + (238 x 1)**, which equals 238,238. The 7-11-13 coincidence—a coincidence no more!

$$238 \times 7 \times 11 \times 13$$
$$= 238 \times 1001$$
$$= (238 \times 1000) + (238 \times 1)$$
$$= 238{,}000 + 238$$
$$= 238238$$

Here's a little bonus for you. Are you ready to discover a similar pattern that works for both two-digit and four-digit numbers? Try multiplying two-digit numbers by 101 **(ab x 101 = abab)** and four-digit numbers by 10,001 **(abcd x 10,001 = abcdabcd)**.

$$52 \times 101$$
$$= (52 \times 100) + (52 \times 1)$$
$$= 5200 + 52$$
$$= 5252$$

$$3579 \times 10001$$
$$= (3579 \times 10000) + (3579 \times 1)$$
$$= 35790000 + 3579$$
$$= 35793579$$

But don't just take my word for it, grab a few numbers and give it a try yourself! You'll be amazed at how easily you can replicate this pattern.

Two-Digit Multiplication Rainbows

My mentor taught me this quick way to multiply two-digit numbers ten years ago and it has stuck with me through all my years as an engineer. Let's start off with easy two-digit numbers and then work our way through large two-digit numbers, where carrying may be needed. With a bit of practice, you'll soon be able to multiply any two-digit number thrown your way!

31 X 12

Steps

1. First, multiply the first two digits in each number **(3 x 1 = 3)**. This will be the first part of your answer.

31 X 12 = 3

2. Next, leave a space in your answer (the middle digit will go here) and move on to the last digit. To calculate this, simply multiply the last two digits in each number **(1 x 2 = 2)**.

31 X 12 = 3 _ 2

Now for the middle digit! This is where we're going to draw a rainbow. First, multiply the two inside digits (the last digit of the first number by the first digit of the second number).

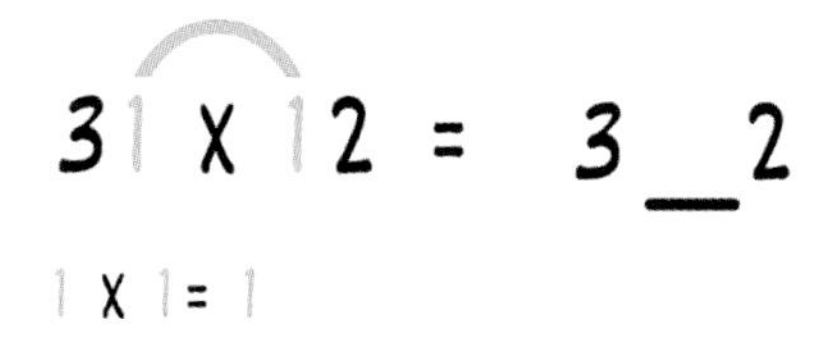

Then, multiply the outside digits (the first digit of the first number and the second digit of the second number).

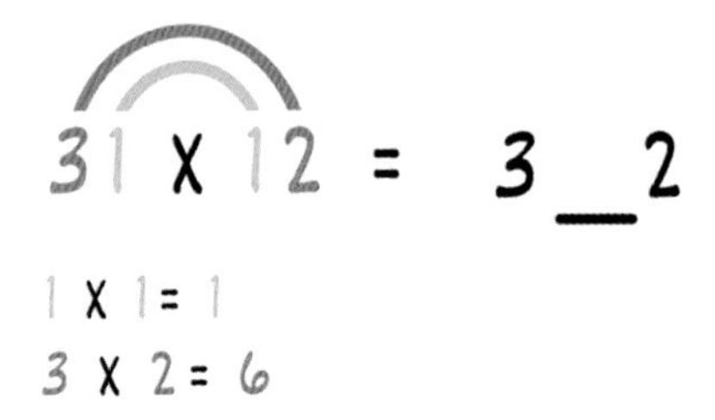

Now add the numbers in steps 3 and 4 to get the middle answer: **1 + 6 = 7**, so **31 x 12 = 372**!

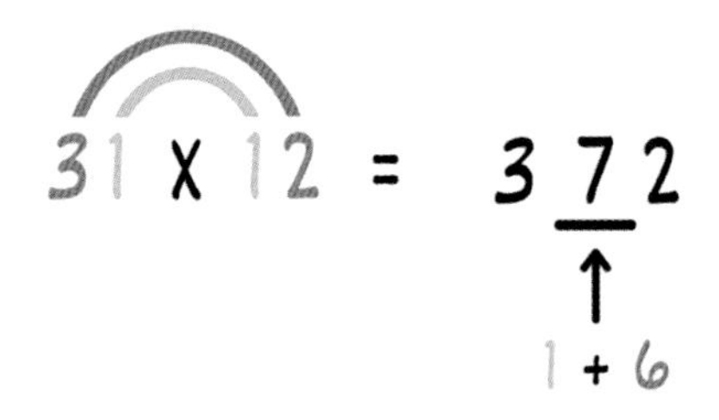

Pretty straightforward, isn't it? Now if your middle digit adds up to 10 or above in step 4, you'll need to do one extra step to carry the tens digit over. Let's try an example where this happens. Let's say you tutor math to younger kids to earn a bit of extra money. If you charge $52 per hour and tutored 81 hours in a year, how much did you earn altogether?

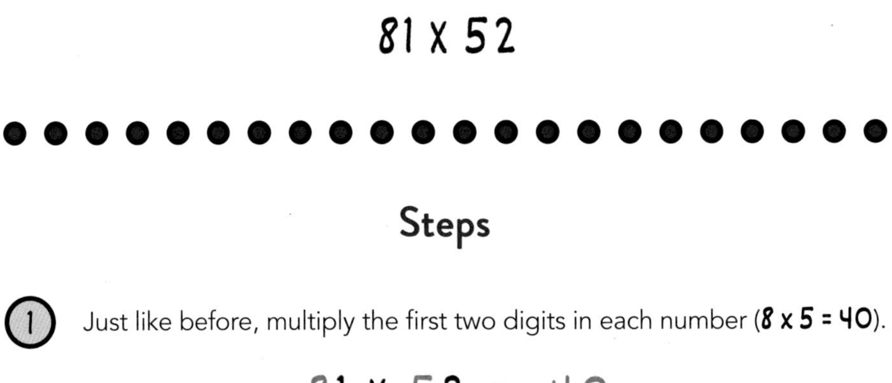

81 X 52

Steps

1. Just like before, multiply the first two digits in each number (**8 x 5 = 40**).

81 X 52 = 40

Then, leave room for the middle digit and calculate the last digit by multiplying the last two digits in each number (**1 x 2 = 2**).

81 X 52 = 40_2

Now let's draw our rainbows. First, multiply the inside digits.

81 X 52 = 40_2

1 X 5 = 5

Then, multiply the outside digits.

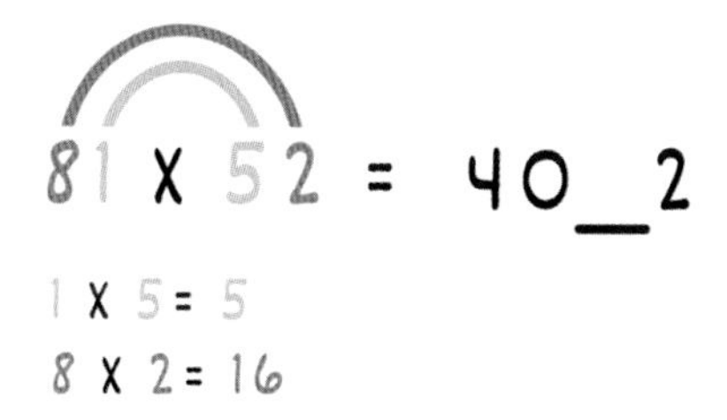

Now add those two numbers (**5 + 16 = 21**). Because 21 is greater than 9, we're going to slot the ones digit (the 1 in 21) into our answer's middle digit and carry the tens digit (the 2 in 21) over to the left.

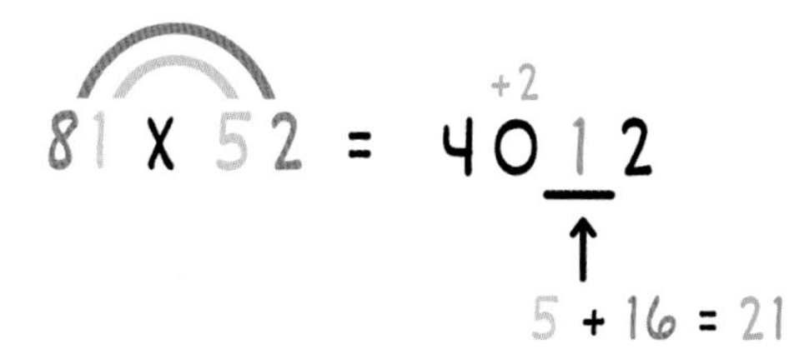

Adding 2 to 0, we get our final answer of 4,212! This means that if you charge $52 an hour for tutoring and tutored a total of 81 hours, you would earn $4,212!

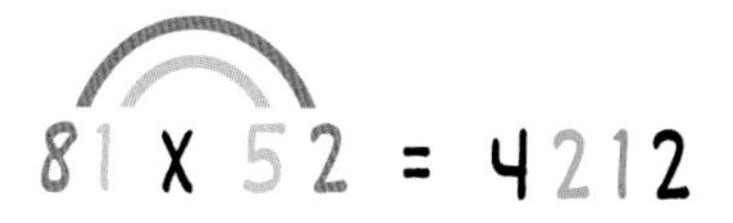

Practice

13 x 21 **42 x 14** **95 x 72**

Why Does This Work?

Let's dissect our first example, **31 x 12**. If we create a 31-unit by 12-unit rectangle, what is the total area (**31 x 12**)?

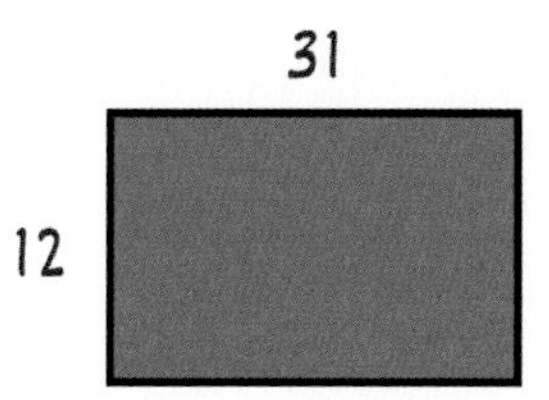

The best way to solve this is by splitting the rectangle into smaller chunks, finding the areas of each small chunk and then adding them together. There are many ways you can split the rectangle, but a great way is by splitting 31 and 12 by their place values (**31 = 30 + 1** and **12 = 10 + 2**).

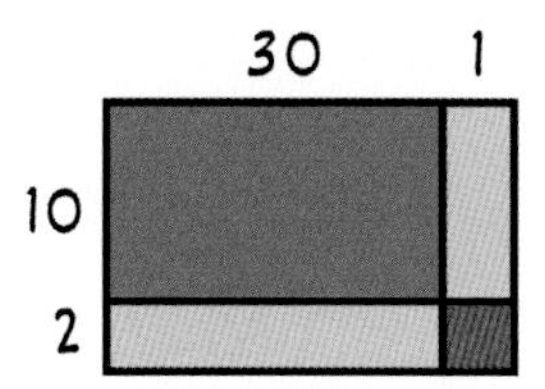

Calculating the area of the smaller rectangles and adding them together, we get our answer of 372! The blue rectangle accounts for the 300 in 372, the red rectangle accounts for the 2 in 372 and the two yellow rectangles together equal 70 in 372.

31 x 12 = 300 + 60 + 10 + 2 = 372

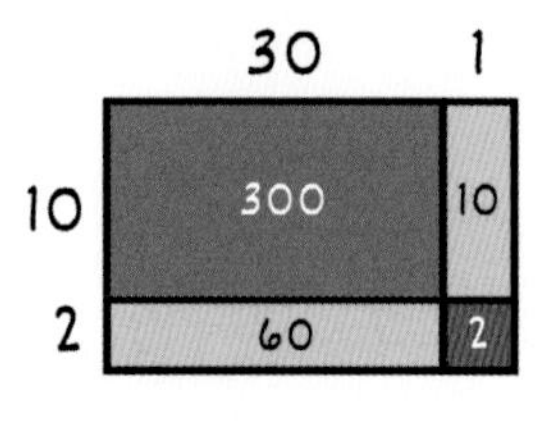

Easy Three-Digit Multiplication

Have you ever played badminton before? I have fond memories of playing this game every day as a kid when I lived in Singapore. It's a fun and exciting game where the first person to score 21 points wins. If you take badminton more seriously, you'll begin to get ranked by the number of points you've scored throughout your life. Let's say a player wins 312 games, which is quite an accomplishment! Assuming no deuces, what would their ranking in points be? Let's find out.

$$312 \times 21$$

Let's solve this by drawing rainbows!

Steps

1. First, multiply the first two digits in each number (**3 x 2 = 6**). This will be the first part of your answer.

$$\underline{3}12 \times \underline{2}1 = 6$$

2. Next, leave two spaces in your answer (the middle two digits will go here) and move on to the last digit. To calculate this, simply multiply the last two digits in each number (**2 x 1 = 2**).

$$31\underline{2} \times 2\underline{1} = 6\_\_2$$

3 Now let's solve for the answer's tens place. Draw a rainbow that multiplies the ones digit of the three-digit number by the tens digit of the two-digit number (**2 x 2 = 4**) and the tens digit of the three-digit number by the ones digit of the two-digit number (**1 x 1 = 1**). Add these together to get the answer's tens place (**4 + 1 = 5**).

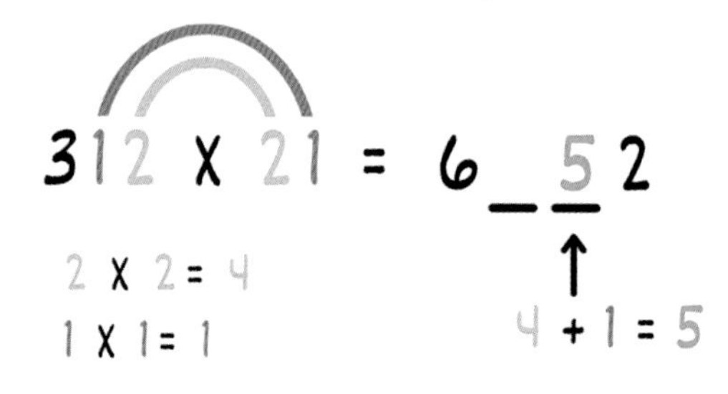

Now for the answer's hundreds place! This time, draw a rainbow below your numbers, multiplying the tens digit of the three-digit number by the tens digit of the two-digit number (**1 x 2 = 2**) and the hundreds digit of the three-digit number by the ones digit of the two-digit number (**3 x 1 = 3**). Adding these together, we get the answer's hundreds place (**3 + 2 = 5**). With that, you've now solved **312 x 21 = 6,552**! This means that for a player who wins 312 games without any deuces, they would have scored a total of 6,552 points!

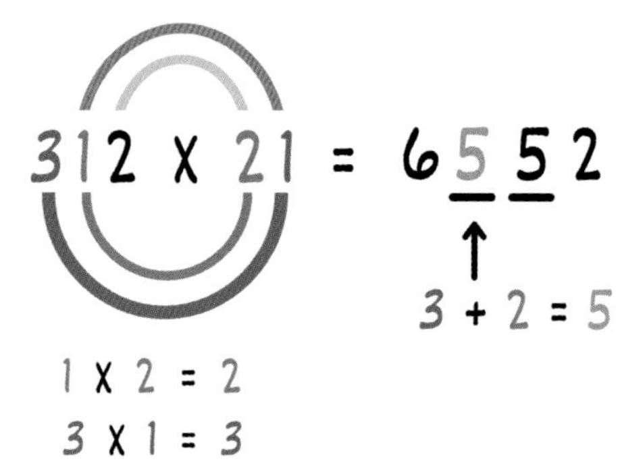

Not too bad, right? Similarly to the previous exercise, if any of your digits are 10 or above, you'll need to carry the tens place over to the next left digit in your answer.

Practice

121 x 31   **821 x 23**   **458 x 72**

Why Does This Work?

We can represent the product of **312 x 21** as the area of a rectangle. However, how would you solve for the area?

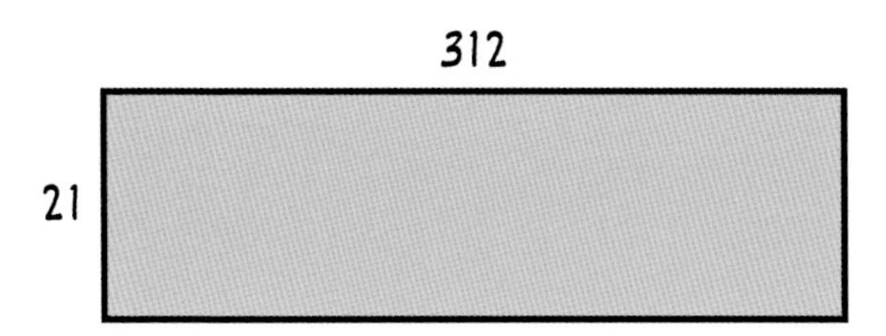

The best way to solve this is by splitting the rectangle into smaller chunks, finding the areas of each small chunk and then adding them together. There are many ways you can split the rectangle, but a great way is by splitting 312 and 21 by their place values (**312 = 300 + 10 + 2** and **21 = 20 + 1**).

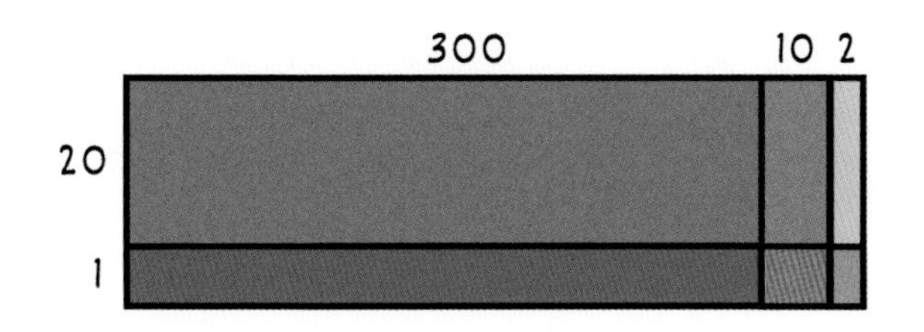

Calculating the area of the smaller rectangles and adding them together, we get our answer of 6,552!

312 X 21 = 6,000 + 300 + 200 + 40 + 10 + 2 = 6,552

Big Numbers? Count Balloons!

Multiplying big numbers can be a chore, but if your numbers end in zeros, you're in luck. Just remember—count your balloons and don't let go!

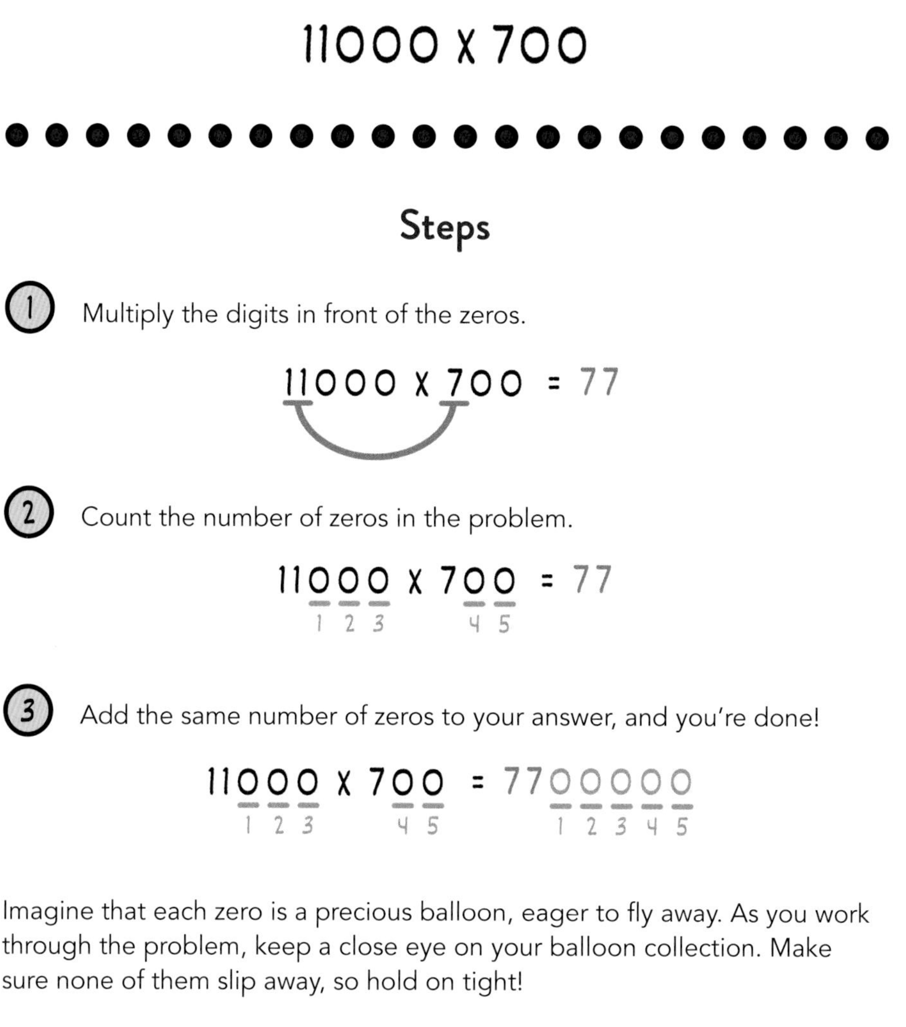

$$11000 \times 700$$

Steps

1. Multiply the digits in front of the zeros.

$$11000 \times 700 = 77$$

2. Count the number of zeros in the problem.

$$11\underset{1}{0}\underset{2}{0}\underset{3}{0} \times 7\underset{4}{0}\underset{5}{0} = 77$$

3. Add the same number of zeros to your answer, and you're done!

$$11\underset{1}{0}\underset{2}{0}\underset{3}{0} \times 7\underset{4}{0}\underset{5}{0} = 77\underset{1}{0}\underset{2}{0}\underset{3}{0}\underset{4}{0}\underset{5}{0}$$

Imagine that each zero is a precious balloon, eager to fly away. As you work through the problem, keep a close eye on your balloon collection. Make sure none of them slip away, so hold on tight!

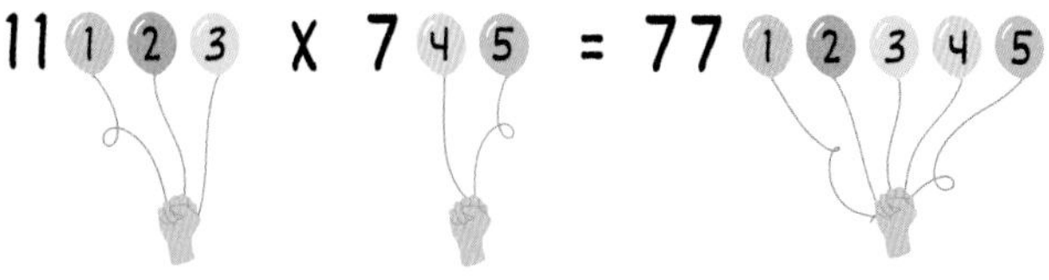

Practice

How many trailing zeros will the answer for **20 x 10,100** have?

10,300 x 20

2,000 x 40 x 700

Why Does This Work?

A number becomes ten times bigger with every trailing zero at the end. For example, 700 is ten times bigger than 70, and 70 is ten times bigger than 7. If we multiply **3 x 2**, we get 6. If we multiply **30 x 2**, we get 60 (which is ten times bigger than **3 x 2**). And if we multiply **300 x 2** or **30 x 20**, we get 600 (which is a hundred times bigger than **3 x 2**). See how that works? It's all about the zeros at the end!

A Guide to Calculating Your Business Earnings

For any current or future business owners, here's an easy way to estimate your earnings. Let's say you have a water bottle business and sold 7,982 bottles for $31 each the past year. How much did you earn? Instead of reaching for a calculator, we can do a quick estimation. Start by rounding your numbers—7,982 to 8,000 and $31 to $30. Then, just multiply 8,000 by $30 using this trick to get $240,000. It's not exactly the actual amount of $247,442, but it's a good estimate and it only took a few seconds to figure out!

8000 X 30 = 240000

When in Doubt, Draw a Box

If you've been setting up all your multiplication problems like this . . .

$$\begin{array}{r} 213 \\ \times 72 \\ \hline \end{array}$$

. . . then it's time to try out the *box method*! This method ditches the traditional procedure—it's all about breaking numbers down and having some fun with them. This is why schools all over the world are now teaching this method, so let's see what it's all about!

$$213 \times 72$$

Steps

1. The first thing you're going to do is draw a box! The size of your box will depend on how many digits are in each number you're multiplying. Because 213 has three digits and 72 has two digits, your box will have two rows and three columns (or three rows and two columns, but I prefer to draw my boxes horizontally).

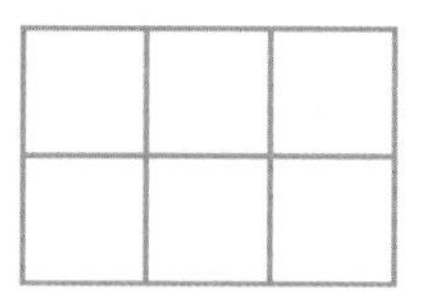

Here are a few more examples of other multiplied numbers and their boxes. Remember, the number of digits in each number will match the number of rows or columns!

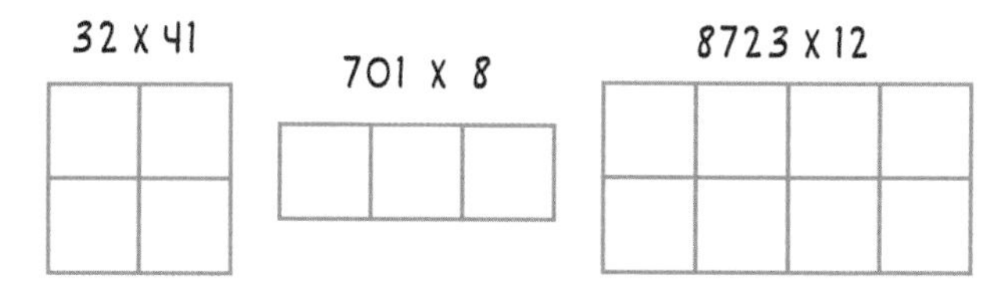

2. Going back to our example, let's break our numbers down into their place values. For example, **213 = 200 + 10 + 3** and **72 = 70 + 2**.

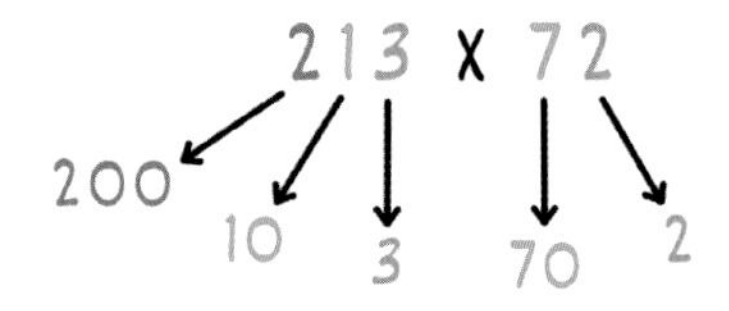

3. Now write these new numbers around the edges of your box. Be sure to line up the correct number of digits with each row and column (200, 10 and 3 will be along the side with three blocks while 70 and 2 will be along the side with two blocks).

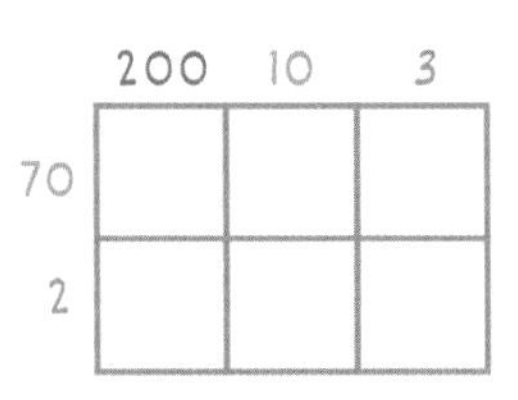

4. Multiply the numbers around each block. To multiply big numbers with trailing zeros (for example, **70 x 200**), simply multiply the digits before the zeros (**7 x 2 = 14**) and then add the same number of zeros back into the answer (because there are a total of three zeros in 70 and 200, add three zeros behind 14 to get 14,000). To learn more about this technique, check out the section Big Numbers? Count Balloons (page 92).

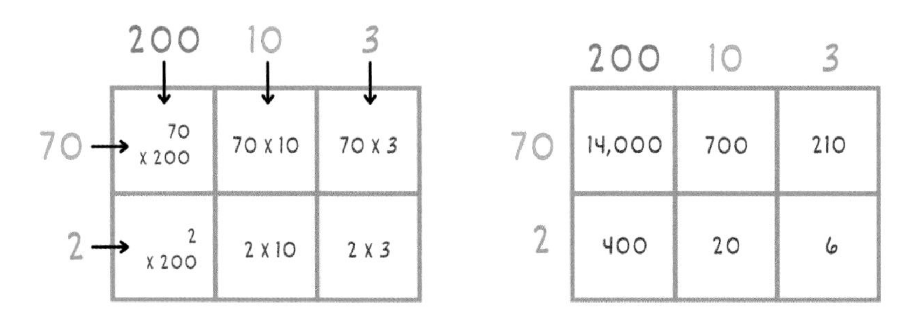

5. Finally, add all the numbers inside your box and that will be the answer: **213 x 72 = 14,000 + 700 + 400 + 210 + 20 + 6 = 15,336**.

$$\begin{array}{r} 14000 \\ 700 \\ 400 \\ +\quad 210 \\ 20 \\ 6 \\ \hline 15{,}336 \end{array}$$

You may be wondering, *"We just went from multiplying two numbers to adding six numbers! Why is this even considered a shortcut?"* Close your eyes for a second and try to multiply **213 x 72** in your head. Not so easy, is it? However, if I asked you to start at 14,000 and then add 700, then 400, then 210, then 20 and then 6, you'll be able to do each step easily. Although we made more moves, it's still easier for our brains to process six easy steps than one big leap!

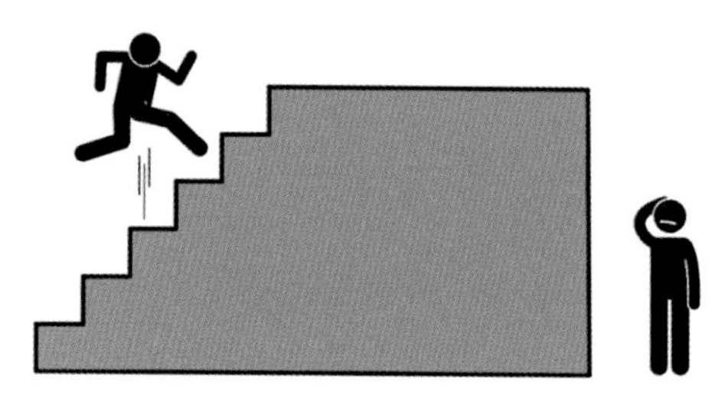

Practice

Draw the box for: **362 x 12,803**

27 x 8,130

123 x 123

Why Does This Work?

The box method is a fantastic way to multiply two numbers, and many schools are starting to prefer it over the traditional method. This is because the box method *actually* explains why multiplication works! All we're doing is breaking our numbers down into their place values, multiplying them together and then adding up the partial products. It's a great way to multiply numbers, whether you're writing it down or doing a mental calculation, because this method breaks the problem down into manageable chunks that are easy to work with!

How Far Have You Traveled?

Whenever you embark on a journey, whether by bike, car, boat or airplane, you can easily estimate the distance you've covered if you know your traveling speed and duration. For instance, when I went to my sister's graduation in Washington, DC, last May, I traveled by plane at an average speed of 530 mph (853 km/h) for 4 hours. How far did I travel? We can quickly find out by multiplying the speed (530 mph) by the duration (4 hours) using the box method!

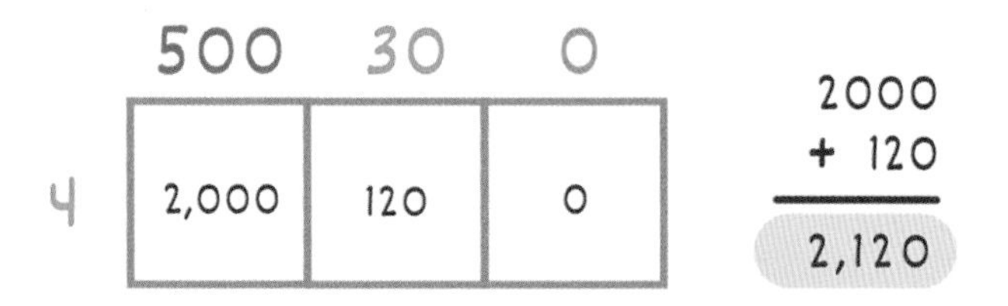

After traveling for 4 hours at an average speed of 530 mph, I traveled a total distance of 2,120 miles (3,412 km)!

Tired of Multiplying? Use Lines!

Ready for a change of scenery from all the numbers? Let's take a break and refresh your mind with this creative trick. I'm going to show you how to multiply numbers in a completely new way—by using lines!

13 X 21

Steps

1. In the next few steps, we're going to draw a set of lines for every digit in our multiplication problem. The first digit in **13 x 21** is 1, so let's draw 1 line at a 45-degree angle going from the bottom left to the top right.

The next digit in **13 x 21** is 3, so this time let's draw 3 lines parallel to the line we drew earlier but shifted toward the bottom right.

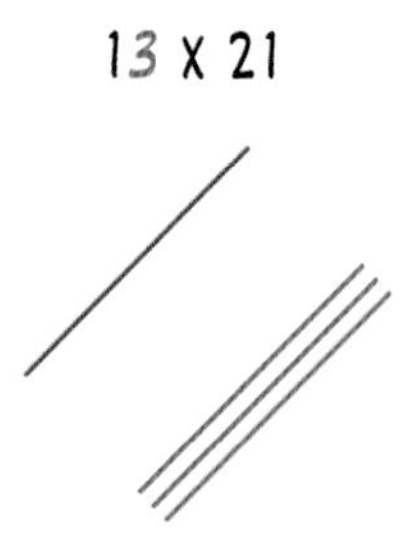

3. The third digit in **13 x 21** is 2. We're going to draw 2 lines but this time perpendicular to the lines we previously drew. Remember, lines for the same number are all parallel but lines for different numbers are perpendicular to each other!

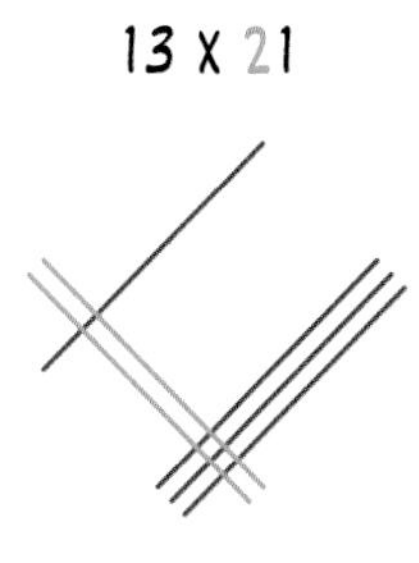

The last digit in **13 x 21** is 1, so let's close off with 1 more line parallel to the 2 lines we last drew. You should now have a square rotated 45 degrees with extra lines on each side.

This is where things get interesting. Do you see where the lines intersect each other? We're going to draw 3 circles that divide them into 3 sections: right, middle and left.

Starting with the right circle, count the number of times lines intersect inside. I count 3 line intersections, so let's write a 3 below our circle.

7 Moving left, count the number of times the lines intersect in the middle circle.

Finally, count the number of times the lines intersect in the left circle.

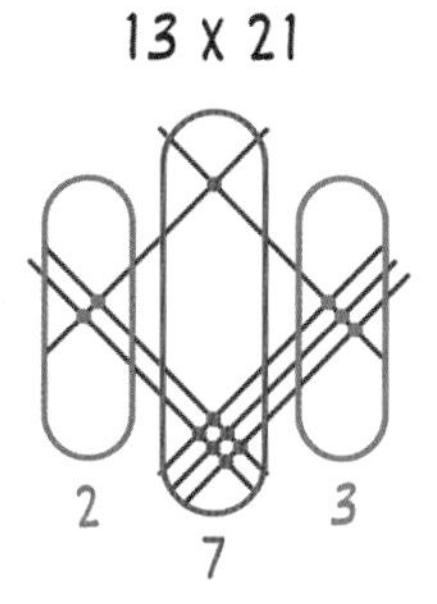

Now combine the three numbers you wrote down from left to right and that will be your final answer!

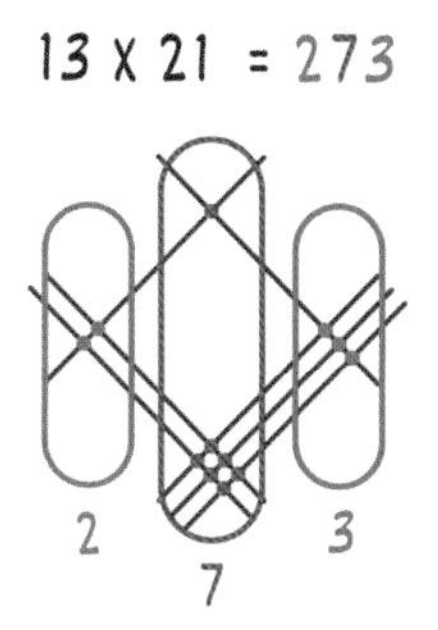

Amazing, isn't it? This trick works for any two-digit number, but if you have 10 or more line intersections in the middle or right circle, you'll need one extra step to get the answer. Let's say you're multiplying **51 x 32**. You've already drawn your lines and circles and are now counting line intersections when you count 13 intersections in the middle circle.

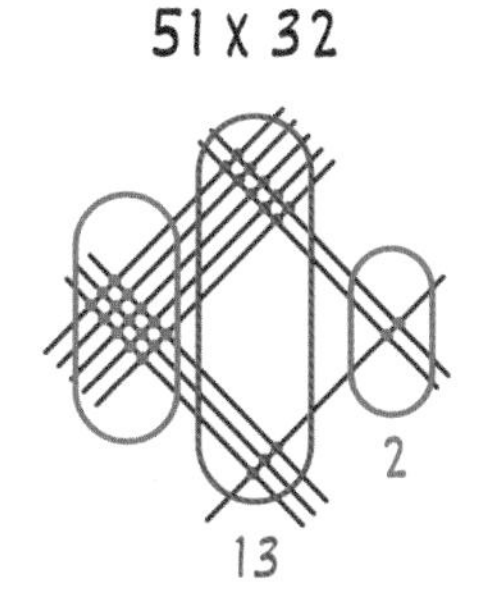

Because 13 is greater than 9, you'll need to do this—keep the 3 in the ones place but carry the 1 in the tens place over to the next circle on the left.

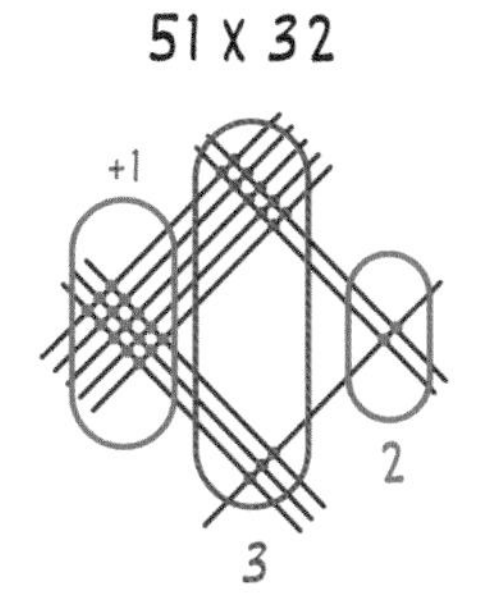

Count the intersections as you would in the left circle but add the 1 from 13 to the final count. There are 15 intersections, but after adding 1 from earlier you'll write down 16. Because there are no more circles to the left, you don't need to carry anything over from here.

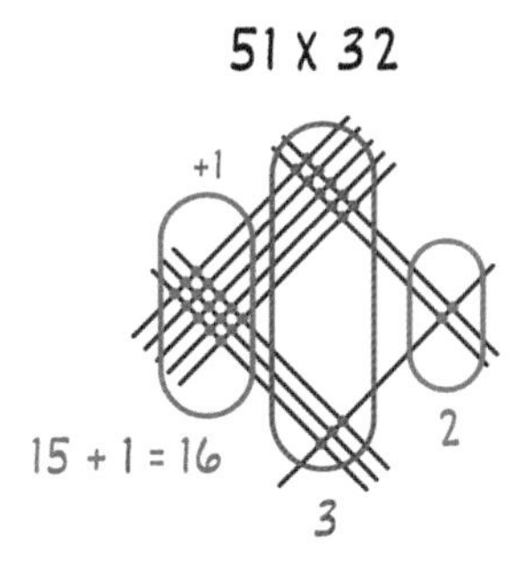

Now you can combine all the numbers from left to right again to get the answer!

51 X 32 = 1632

Practice

62 x 13

132 x 21

122 x 122

Why Does This Work?

This may look like math sorcery, but the explanation is simple—the number of line intersections you count will always equal the product of how many lines intersect. For example, if we have 5 lines intersecting 8 lines, we will have 40 intersections because **5 x 8 = 40**.

Going back to our first example, all we're doing is first multiplying the tens digits in 13 and 21 to solve for the answer's hundreds place value.

$$13 \times 21 = 2$$

Then, we multiply the ones digits to solve for the answer's ones place value.

$$13 \times 21 = 2\_3$$

And finally, we multiply the inner and outer digits and add their products to get the answer's tens place value. To learn more about how we did this, flip to the section Two-Digit Multiplication Rainbows (page 85).

$$13 \times 21 = 273$$

$$3 \times 2 + 1 \times 1$$

Division? No Sweat!

Doing division used to give me a headache, especially when dealing with large numbers, decimals and don't even get me started on long division! Luckily, there are a few simple tricks you can use to make division much easier. In this chapter, we'll explore some of the best division tricks you can use every day to save time and reduce stress. But before we dive in, let's start with a quick refresher on some basic principles of division.

In the last section, we learned that multiplication is just a game of repeated addition. For example, **5 x 2** is the same as adding 2 together 5 times.

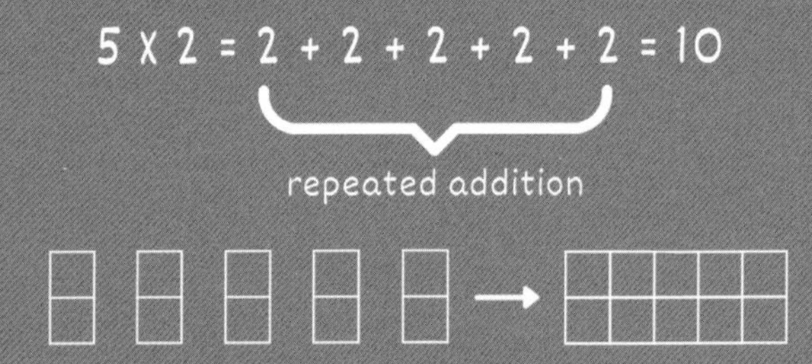

With that in mind, what is division? It's just a game of repeated subtraction! Let's take **10 ÷ 2** for example—to find the answer, subtract groups of 2 from 10 until you reach the ultimate goal: zero. And how many times did you subtract 2? Exactly 5 times! This is why **10 ÷ 2 = 5**.

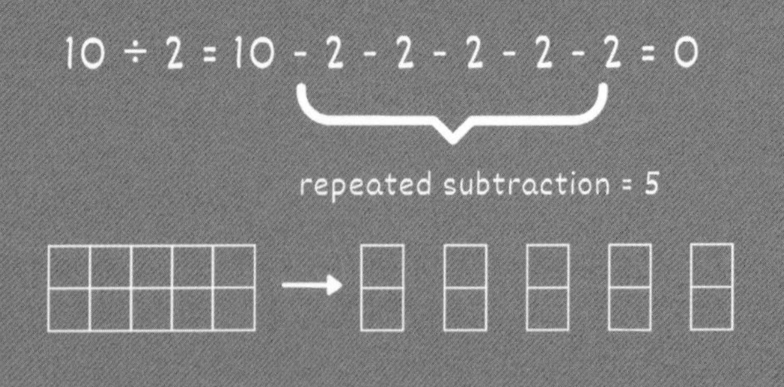

But like in any game, things don't always go as planned and sometimes we don't end up with a perfect zero. That's when we get a remainder. A remainder will always be smaller than the number you're dividing by, and it's the reason behind the existence of fractions and decimals. Something extra to spice up the math world.

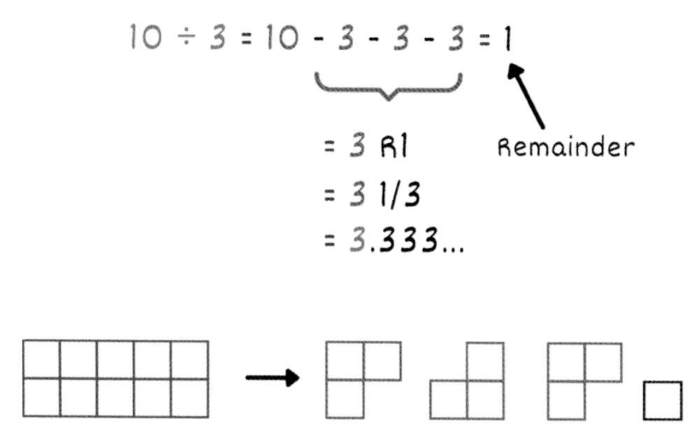

What happens if you divide a number by a fraction like ½? It's the same concept—simply break each unit into ½ pieces and keep subtracting groups of ½ until you reach zero. It's like cutting pieces of cake into even smaller pieces.

$$10 \div \frac{1}{2} = 10 - \frac{1}{2} - \frac{1}{2} - \frac{1}{2} - \frac{1}{2} - \frac{1}{2} - \frac{1}{2} - \frac{1}{2} - \frac{1}{2} - \frac{1}{2} - \frac{1}{2}$$

$$- \frac{1}{2} - \frac{1}{2} - \frac{1}{2} - \frac{1}{2} - \frac{1}{2} - \frac{1}{2} - \frac{1}{2} - \frac{1}{2} - \frac{1}{2} - \frac{1}{2} = 0$$

$$= 20$$

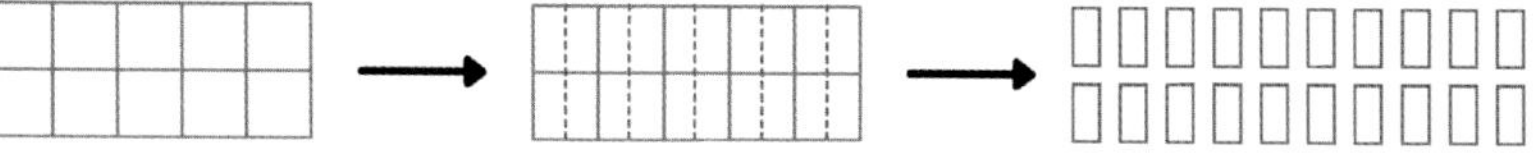

Can you divide a number by zero? This has stumped people all over the world because if you type it into your calculator, you'll get a message that says ERROR or UNDEFINED. But do you know why it's impossible? Let's put our game of repeated subtraction to the test and divide 10 by 0!

$$10 \div 0 = 10 - 0 - 0 - 0 - 0 - 0 - 0 - 0 - 0 - 0 - 0$$
$$- 0 - 0 - 0 - 0 - 0 - 0 - 0 - 0 - 0 - 0$$
$$- 0 - 0 - 0 - 0 - 0 - 0 - 0 - 0 - 0 - 0 = \text{undefined}$$

No matter how many times we subtract zero from 10, the 10 will never go down to zero. It's like trying to scoop water from a bowl using a fork; it just doesn't work! That's why $10 \div 0$ = UNDEFINED.

One last thing before we dive into our division tricks. Dividing only works when you break things down into *equal* pieces. Take a look at these two cakes. Although they are both *cut* into 8 pieces, only the second cake is *divided equally* into 8 pieces.

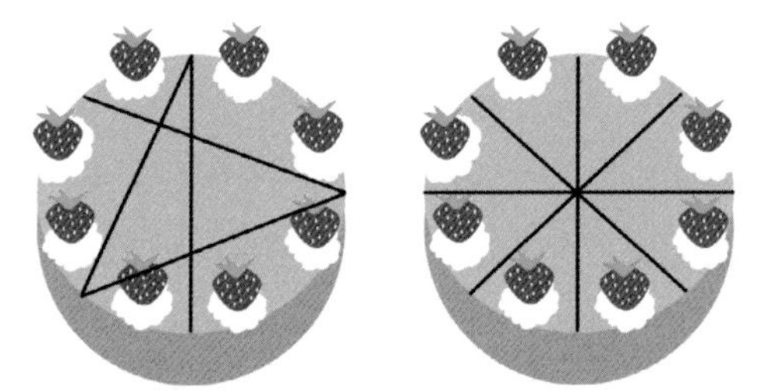

Here's a fun brain teaser for you: How can you divide the cake into 8 equal pieces using only 3 straight slices? Give it a try and see if you can come up with a solution!

If this is stumping you, check out the answers in the back of this book for a helpful hint (page 199). Now that you're an expert on division principles, let's dive into our handy tricks. Say goodbye to division headaches once and for all!

Instantly Divide by 5 (and 0.5, 50, 500)

We can all agree that dividing by 5 is easy when the numbers end in 5 or 0, like **20 ÷ 5** or **35 ÷ 5**. But what about other numbers like **43 ÷ 5**, **91 ÷ 5** or **807 ÷ 5**? Here's a crowd-favorite trick that will make this super easy! Let's take **43 ÷ 5** as an example.

$$43 \div 5$$

Steps

1. First, double your number.

$$43 \times 2 = 86$$

2. Then, divide by 10 and that will be your answer! To divide anything by 10, simply move the decimal point once to the left.

$$86 \div 10 = 8.6$$

Now here's a challenge for you—how would you apply this logic to divide 43 by 0.5, 50 or 500? Let's take a look.

43 ÷ 0.5
To divide 43 by 0.5, all you need to do is multiply 43 by 2. So, **43 x 2 = 86**, and thus **43 ÷ 0.5 = 86**.

43 ÷ 50
To divide 43 by 50, first multiply 43 by 2 (**43 x 2 = 86**) and then divide by 100 (**86 ÷ 100 = 0.86**).

43 ÷ 500
To divide 43 by 500, first multiply 43 by 2 (**43 x 2 = 86**) and then divide by 1,000 (**86 ÷ 1,000 = 0.086**).

Practice

32 ÷ 5   231 ÷ 50   4,200 ÷ 500

Why Does This Work?

The easiest numbers for your brain to mentally work with are 0, 1, 2, 10 and 100. Because **5 = 10 ÷ 2**, dividing a number by 5 is the same as multiplying it by 2 and then dividing it by 10. It will always be easier to solve two steps using 0, 1, 2, 10 and 100 than one step using difficult numbers!

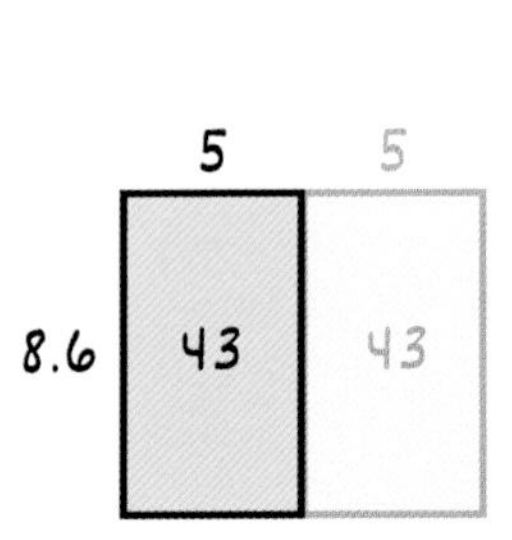

How to Make the Most Out of Your Time

Basketball has five key positions: point guard, shooting guard, small forward, power forward and center.

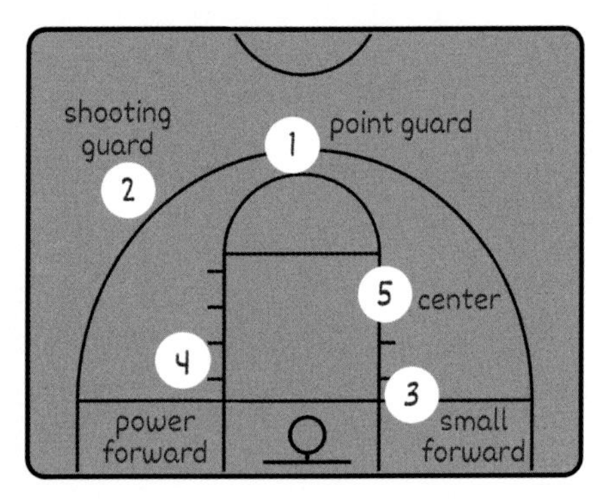

Tomorrow I'll have 1.5 hours (or 90 minutes) to practice basketball. If I want to spend an equal amount of time perfecting my skills for each of the five positions, how much time should I dedicate to each one? Rather than trying to figure out what **90 ÷ 5** is, we can instead multiply 90 by 2 (**90 x 2 = 180**) and then divide by 10 (**180 ÷ 10 = 18**). This means I have 18 minutes to practice each skill!

Can You Divide by 25 (0.25, 2.5, 250)?

Have you used the Pomodoro method to study before? It's a famous time management method named after those popular tomato timers from the 1980s that can make even the most daunting task feel doable!

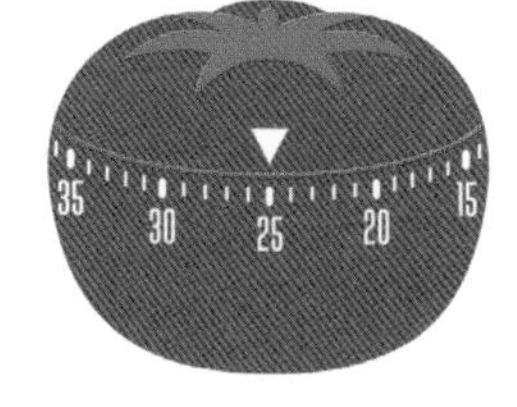

The concept is simple—you study for 25 minutes, followed by a 5-minute break. You then repeat this pattern until you finish studying. For example, if you study for 120 minutes, you will have four 25-minute study intervals with breaks in between. However, if you skip the breaks, how many 25-minute study intervals can you complete within 120 minutes? Here's an easy way to calculate this in your head!

$$120 \div 25$$

Steps

1. First, multiply the dividend by 4. The dividend refers to the number being divided, which in this case is 120.

$$120 \times 4 = 480$$

2. Then, divide by 100 and that will be your answer! To divide anything by 100, simply move the decimal point twice to the left. Therefore, $120 \div 25 = 4.8$.

$$480 \div 100 = 4.8$$

This means that if you need to study for 120 minutes straight without taking any breaks, you would have 4.8 25-minute study intervals. Nice! Now using the same logic as the previous steps, how would you divide numbers by 0.25, 2.5 and 250? Let's take a look.

$21 \div 0.25$

To divide 21 by 0.25, all you need to do is multiply 21 by 4, so **$21 \times 4 = 84$** and thus **$21 \div 0.25 = 84$**.

$121 \div 2.5$

To divide 121 by 2.5, first multiply 121 by 4 (**$121 \times 4 = 484$**) and then divide by 10 (**$484 \div 10 = 48.4$**).

$702 \div 250$

To divide 702 by 250, first multiply 702 by 4 (**$702 \times 4 = 2{,}808$**) and then divide by 1,000 (**$2{,}808 \div 1{,}000 = 2.808$**).

Practice

$112 \div 25$ $1 \div 2.5$ $90 \div 250$

Why Does This Work?

Would you agree with me that **$25 = 100 \div 4$**? Therefore, dividing a number by 25 is the same as multiplying it by 4 and then dividing it by 100. It is much easier for our brains to work with 4 and 100, which is why dividing by 25 becomes easier when we break it down!

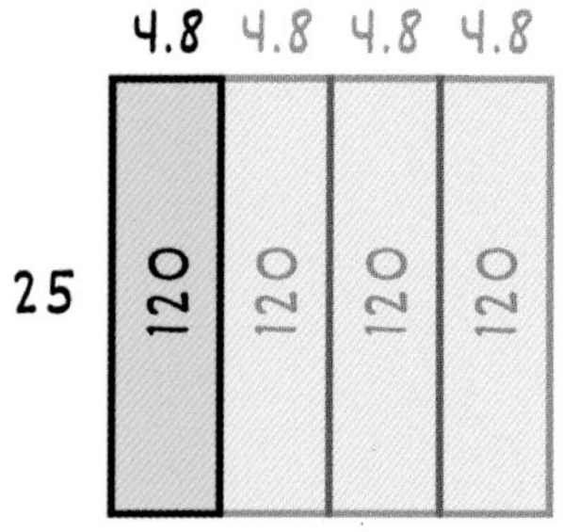

4.8 4.8 4.8 4.8
25
120 120 120 120

Divide by 1.25 (0.125, 12.5, 125) in Your Head

Do you use the playback speed adjustment feature on YouTube videos? I have to admit, I often use it to speed up videos to play at 1.25X the speed when the speaker is talking a little too slow.

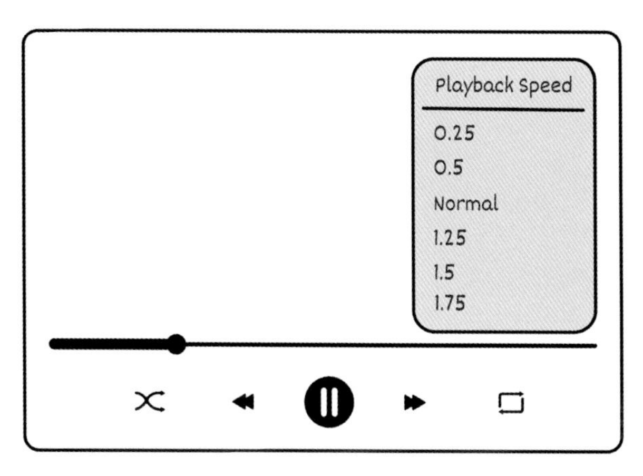

The other day I caught myself making this calculation—if I'm watching a 7-minute video at 1.25X the speed, how much time will I save?

$$7 \div 1.25$$

Steps

 First, multiply your number by 8.

$$7 \times 8 = 56$$

2. Then, divide by 10 and that will be your answer! A 7-minute video will only take 5.6 minutes to watch it at 1.25X the speed.

$$56 \div 10 = 5.6$$

Using this logic, how would you divide numbers by 0.125, 12.5 and 125?

1 ÷ 0.125

To divide 1 by 0.125, all you need to do is multiply 1 by 8. So, **1 x 8 = 8**, and thus **1 ÷ 0.125 = 8**. Why do we not divide by 10 here? Well, what does 0.125 equal to as a fraction? It's **⅛**! And dividing by a fraction (**1 ÷ ⅛**) is the same as multiplying by its reciprocal (**1 x 8**)! If you need a quick refresher on reciprocals, a reciprocal is the inverse of a number, obtained by dividing 1 by that number. To find the reciprocal of a fraction, you can simply flip it, interchanging its numerator and denominator.

90 ÷ 12.5

To divide 90 by 12.5, first multiply 90 by 8 (**90 x 8 = 720**) and then divide by 100 (**720 ÷ 100 = 7.2**).

400 ÷ 125

To divide 400 by 125, first multiply 400 by 8 (**400 x 8 = 3,200**) and then divide by 1,000 (**3,200 ÷ 1,000 = 3.2**).

If it's too challenging to multiply a number by 8, try breaking the problem down! Multiplying by 8 is the same as multiplying by 2 three times, and doubling something is always doable. For example, multiplying 400 by 8 is the same as doubling 400 to 800, then 1,600 and finally 3,200.

Practice

100 ÷ 125

25 ÷ 0.125

30 ÷ 12.5

Why Does This Work?

Let's take a closer look at dividing a number by 125. What is the relationship between 125 and 1,000? Well, **1,000 = 125 x 8**! Therefore, dividing a number by 125 is the same as multiplying it by 8 and then dividing it by 1,000. Although 1,000 is a huge number, it is easy for us to divide by it—just move the decimal point three places to the left!

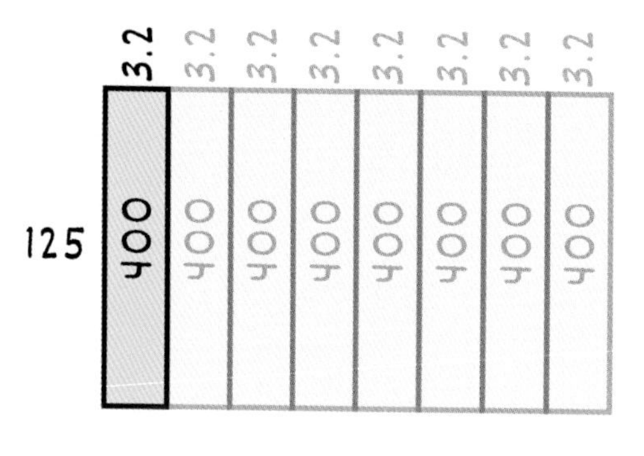

Don't Love Long Division? Try This!

Let's admit it—long division is useful but painful at times. It becomes difficult to do when your divisor (the number you're dividing by) is large. For example, how would you solve this using long division?

$$87\overline{)783}$$

Did you find yourself asking how many times 87 fits into 783? Doesn't this defeat the whole purpose of using long division? The good news is that there is an alternative method you may use to solve this called the *partial quotients method*. The best part? You don't even need to know any of your times tables except for the easy ones like your 2s and 10s.

Let's try the partial quotients method on an easier example problem and divide 173 by 13.

$$13\overline{)173}$$

Steps

1. Set up your problem as if you're doing long division but extend a line all the way down on the right. We're going to spend the next few steps trying to get our dividend (173) down to 0 by subtracting multiples of our divisor (13) from it.

$$13\overline{)173}$$

Focus on the 173 and ask yourself, "What is an easy number I can multiply 13 by that is below 173?" I don't have the 13 times table memorized, but I do know that **13 x 10 = 130** and **13 x 2 = 26**.

```
13 | 173 |
         |    13 x 10 = 130
         |    13 x 2 = 26
         |
```

3

You need to pick one of these numbers to subtract from 173, but which should you choose—130 (**13 x 10**) or 26 (**13 x 2**)? You want to pick the highest multiple of 13 that is still below 173, so let's go with 130! If we picked 26 instead, we would eventually get to the same answer, but by using more steps.

```
13 | 173 |
    -130 | 10
    ----
      43 |
         |
```

4

Repeat the process again and ask yourself, "What is an easy number I can multiply 13 by that is still below 43?" This time, 130 (**13 x 10**) is too high, so let's go with 26 (**13 x 2**). Subtracting 26 from 43 gives us 17. We're getting close!

```
13 | 173 |
    -130 | 10
    ----       13 x 10 = 130
      43 |
     -26 | 2   13 x 2 = 26
     ---
      17 |
```

5 Finally, ask yourself, "What is an easy number I can multiply 13 by that is still below 17?" How about **13 x 1**? Let's write 13 on the left, 1 on the right, and subtract 13 from 17 to get 4.

```
13 | 173
    -130 | 10
    ----
      43         13 X 10 = 130
     -26 | 2
    ----         13 X 2 = 26
      17
     -13 | 1     13 X 1 = 13
    ----
       4
```

When the number at the bottom (4) is less than your divisor (13), you're done! Because we didn't end in a 0, we will have a remainder of 4. To get the final answer, add all the numbers along the right side (**10 + 2 + 1 = 13**). Your final answer for **173 ÷ 13** will be 13 R4, where *R* stands for remainder! To convert the remainder into a fraction, put the remainder (4) over the divisor (13) to get a final answer of 13 4⁄13.

```
13 | 173
    -130 | 10        10
    ----
      43          +   2
     -26 | 2
    ----              1
      17          -----
     -13 | 1         13
    ----
       4
```

$$173 \div 13 = 13\ R4 = 13\frac{4}{13}$$

The beauty of the partial quotients method is that there can be not one, but many ways to get to the answer. For example, instead of subtracting **13 x 2** and then **13 x 1**, you could have subtracted **13 x 1** three times.

```
13 | 173
    -130 | 10
    ----
      43
     -13 | 1
    ----
      30
     -13 | 1
    ----
      17
     -13 | 1
    ----
       4
```

Or you could even just keep on subtracting **13 x 1** from 173 until you reach the answer. It may take you a long time to solve it this way, but it will still work!

```
13 | 173
     -13 | 1
     160
     -13 | 1
     147
     -13 | 1
     134
     -13 | 1
     121
     -13 | 1
     108
     -13 | 1
      95
     -13 | 1
      82
     -13 | 1
      69
     -13 | 1
      56
     -13 | 1
      43
     -13 | 1
      30
     -13 | 1
      17
     -13 | 1
       4
```

There are many ways you can use the partial quotients method, and it's completely up to you which path to take!

Practice

82 ÷ 15

723 ÷ 80

850 ÷ 110

Why Does This Work?

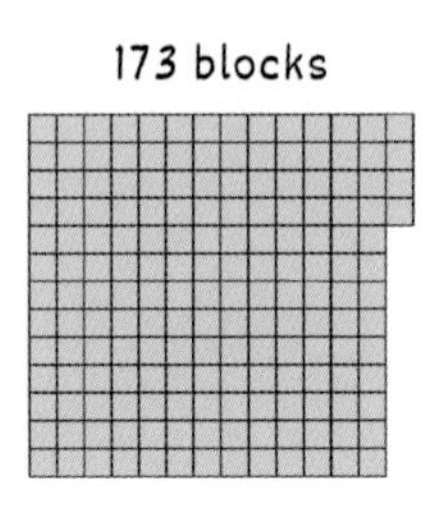

Remember when we established that division is just repeated subtraction until 0? For example, if we break **80 ÷ 10** down, we get **80 - 10 - 10 - 10 - 10 - 10 - 10 - 10 - 10 = 0**. How many times did we subtract 10 from 80? It was 8 times! So **80 ÷ 10 = 8**. In our earlier example, we started with 173 blocks and divided it into groups of 13 blocks, like so:

173 ÷ 13 = 173 - 13 - 13 - 13 - 13 - 13 - 13 - 13 - 13 - 13 - 13 - 13 - 13 - 13 = 4. How many times did we subtract 13? It was 13 times! However, we couldn't get down to a perfect 0, so **173 ÷ 13 = 13 R4**.

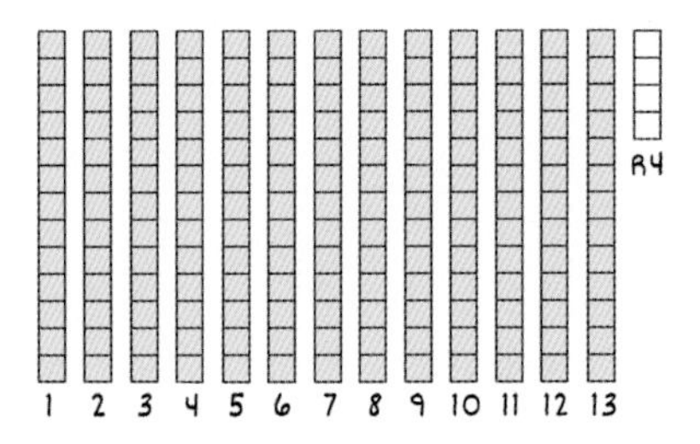

Subtracting 13 thirteen times from 173 is tedious. What if we subtracted multiples of 13 instead? For example, **173 - (10 x 13) - (2 x 13) - (1 x 13) = 4**. At the end of the day, we still subtracted thirteen 13s, but this time only subtracted three groups of them. This is what the partial quotients method is all about!

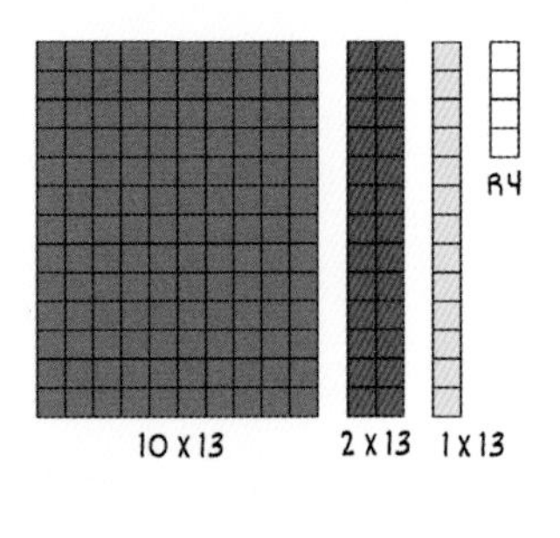

Predict Whether a Number Is Divisible from 2 to 10

Have you ever been to a hackathon before? It's where programmers get together in teams for 24 hours, skip sleeping and work together to build a new software. It's like a marathon but for coding!

Let's say you are planning a hackathon in a conference room that can fit a maximum of 252 programmers. If each team has 6 programmers, is it possible to fit 252 programmers perfectly in teams of 6?

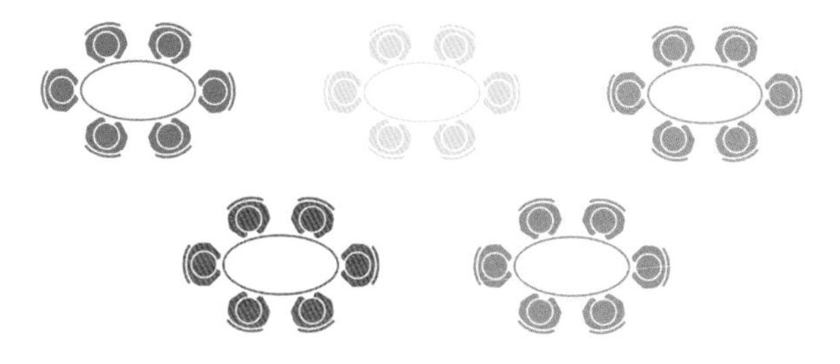

Instead of guessing or doing any calculations, try using these divisibility shortcuts! Keep this handy—this will tell you whether any number is divisible by 2 all the way through 10.

A number is divisible by:

2 if the last digit is an even number (0, 2, 4, 6 or 8)

3 if the sum of all its digits is divisible by 3

4 if the last two digits are divisible by 4

5 if it ends in a 5 or 0

6 if it is divisible by both 2 and 3

7 if when you double the last digit and subtract the sum from the rest of the number, you end up with a number divisible by 7 or get 0

8 if the last three digits are divisible by 8

9 if the sum of all its digits is divisible by 9

10 if it ends in a 0

Let's go through each of these with an example!

Is 1512 divisible by.....

2 ✓

Since the last digit is an even number (2), then 1,512 is divisible by 2!

3 ✓

Let's separate the four digits in 1,512 and add them together: **1 + 5 + 1 + 2 = 9**. Can you divide 9 by 3? Yes! This means that 1,512 is also divisible by 3.

4 ✓

The last two digits in 1,512 are 12. Is 12 divisible by 4? Yes (**12 ÷ 4 = 3**). Therefore, 1,512 is also divisible by 4.

5 ✖

Because 1,512 does not end in a 5 or 0, it is not divisible by 5.

6 ✓

Is 1,512 divisible by both 2 and 3? Yes, we just solved for these, so 1,512 is also divisible by 6!

7 ✓

If we double the last digit in 1,512, we get a 4. Subtracting 4 from the rest of the numbers excluding the 2 in 1,512, we get **151 - 4 = 147**. Now is 147 divisible by 7? You can either use long division for this or repeat this process! Doubling the last number in 147, we get 14. Subtracting 14 from the rest of 147, we get **14 - 14 = 0**. Is 0 divisible by 7? Yes, it is! 0 divided by 7 is just 0. This means that 1,512 is also divisible by 7.

8 ✓

The last three digits in 1,512 are 512. Is 512 divisible by 8? You may need to use long division for this one (or my previous trick on page 114!), but **512 ÷ 8 = 64**. Because 512 is divisible by 8, 1,512 will also be divisible by 8!

9 ✓

Let's separate the four digits in 1,512 and add them together: **1 + 5 + 1 + 2 = 9**. Can you divide 9 by 9? Yes! This means that 1,512 is also divisible by 9.

10 ✖

Because 1,512 does not end in a 0, it is not divisible by 10.

You just determined that 1,512 is divisible by 2, 3, 4, 6, 7, 8 and 9. Once you practice these a little, you'll be able to determine these in seconds!

Practice

Are the following numbers divisible by 2, 3, 4, 5, 6, 7, 8, 9 and 10?

78

864

5,040

Making Perfect Teams

Going back to the hackathon with 252 programmers, would you be able to fit 252 programmers perfectly in teams of 6? Using our handy shortcut, we know that 252 is divisible by 2 because the last digit is even. We also know it's divisible by 3 because the sum of the digits is divisible by 3 (**2 + 5 + 2 = 9**). Therefore, 252 is also divisible by 6 and you will be able to perfectly fit groups of 6 programmers in the room!

The Power of Percentages

What is 23% of 50?

How do I tip 18% on a $37 restaurant bill?

What is a 6% annual interest on a $12,000 loan?

Do these questions give you a headache? If so, then it's time for us to put an end to that. In this section you'll not only learn how to solve these but also do it in your head.

But first, let's jog your memory. What is a percent? A percent is a fancy way of expressing a fraction whose denominator is 100. So, 1% is the same as $\frac{1}{100}$, 37% is the same as $\frac{37}{100}$ and 100% is the same as $\frac{100}{100}$, which is 1.

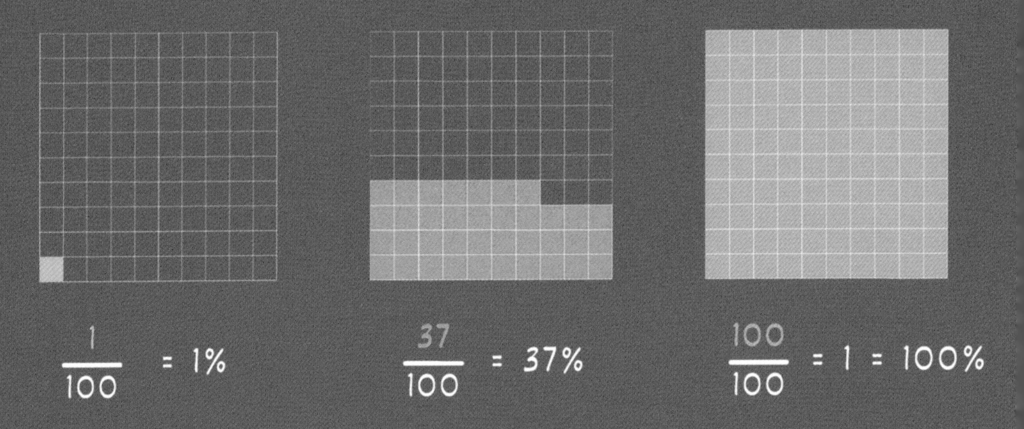

How about a fraction whose denominator is not 100, like $\frac{7}{20}$? It can still be a percentage—simply convert the fraction into an equivalent fraction over 100 or divide the top by the bottom number. Here are four ways you can do it!

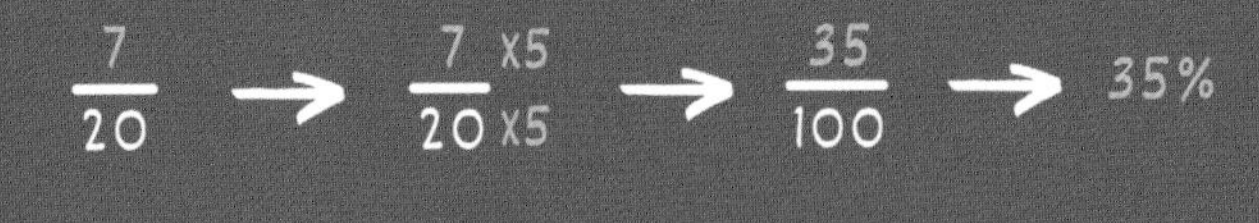

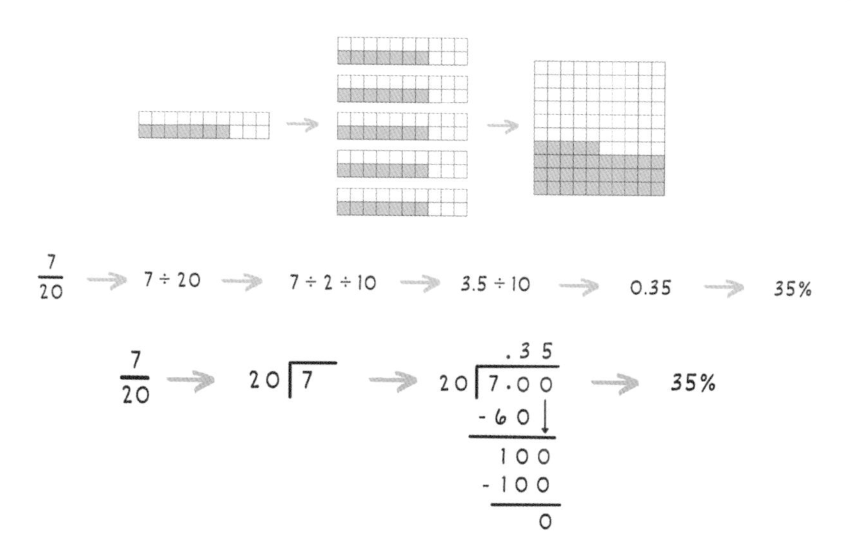

Here's a tip you may find useful when working with percentages. Let's take a closer look at the word *percent*. *Per* means "for each" in Latin and *cent* means "100." Putting them together, *percent* means "for each 100."

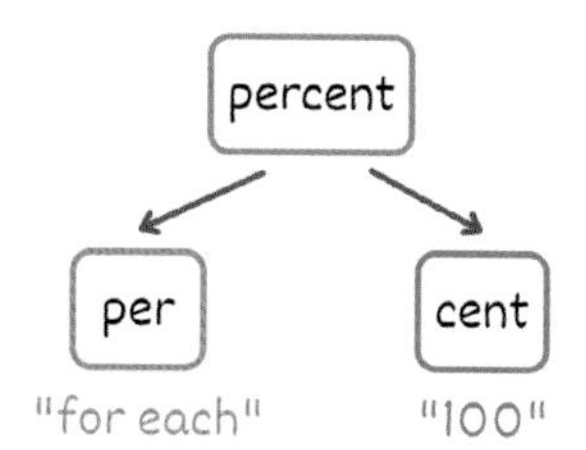

The Latin root *cent* explains why there are 100 *cents* in $1, 100 years in a *cent*ury and 100 *centi*meters in a meter. It's all connected, like a big web of 100s.

1 dollar = 100 X ¢

1 century = 100 X

1 meter = 100 X 1CM

Now that you know how percentages work, let's explore some tricks to solve them headache-free!

Stuck? Reverse Your Percentages

Are you ready for a trick that will change the way you do math forever? Introducing . . . reversing percentages. Doing this will transform difficult percentage problems into simpler ones that you can solve in seconds!

$$16\% \text{ of } 50$$

Steps

1. Reverse your percentages by switching which number you're taking the percentage of.

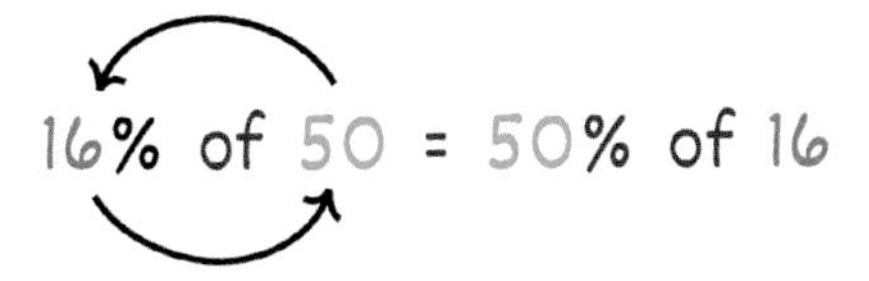

2. Solve your new reversed expression and you're done! So, 50% of 16 is just half of 16, which is 8!

$$50\% \text{ of } 16 = \frac{1}{2} \times 16 = 8$$

Did that just blow your mind? Let's try another one: What is 36% of 25? This looks difficult to solve, but 36% of 25 is the same as 25% of 36, which is just **$36 \div 4 = 9$**. Nice!

This trick is a game changer, but it does have its limitations. Not every percentage problem is made easier by reversing its percentages. For example, reversing 28% of 73 to be 73% of 28 did not make the problem significantly easier to solve!

$$28\% \text{ of } 73 = 73\% \text{ of } 28 = \text{???}$$

Reversing percentages works best when you're finding the percentage of whole numbers like 10, 20, 25, 50, 75 and 100.

Practice

23% of 200 **15% of 20** **12% of 75**

Why Does This Work?

Let's try solving 16% of 50 without reversing percentages. First multiply **16% x 50** and then convert everything to fractions (you may also solve this by converting 16% into the decimal 0.16, but for this example I'm going to use fractions).

$$16\% \times 50 = \frac{16}{100} \times \frac{50}{1} = \frac{16 \times 50}{100 \times 1}$$

Now do you remember the commutative property of multiplication? The order of multiplying numbers does not matter. For example, **3 x 5 = 5 x 3**. Why is this important? Because we can apply this to our current problem and switch the **16 x 50** in the numerator to **50 x 16**.

$$\frac{16 \times 50}{100 \times 1} = \frac{50 \times 16}{100 \times 1}$$

Working backward, let's separate out the fractions, convert them back into percentages and ta-da! You've just proved that 16% of 50 = 50% of 16.

$$\frac{50 \times 16}{100 \times 1} = \frac{50}{100} \times \frac{16}{1} = 50\% \text{ of } 16$$

A Tip on Tipping

You're indulging in a delicious meal and the cost comes out to be $49.28. You would like to leave an 18% tip, but how would you estimate that on the spot? To make things easy, first round your bill to $50 and then take 18% of 50!

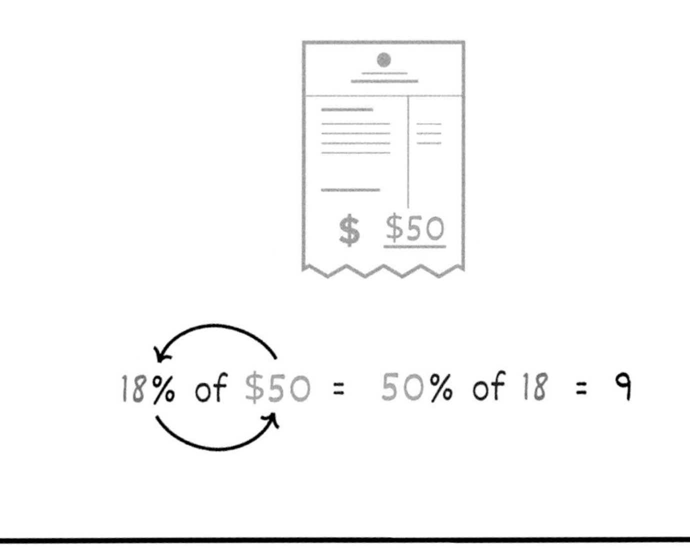

Percentages Made Easy!

Want to learn how to solve problems like these with the snap of a finger?

70% of 20

40% of 60

30% of 90

Take a close look at these three problems. What do you observe about them? All the numbers and percentages are two-digit multiples of 10. When this happens, all you need is one simple step to solve the problem! Let's try solving 70% of 40.

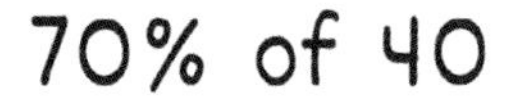

Steps

Take the first digit of each number and multiply them together. It's as easy as that!

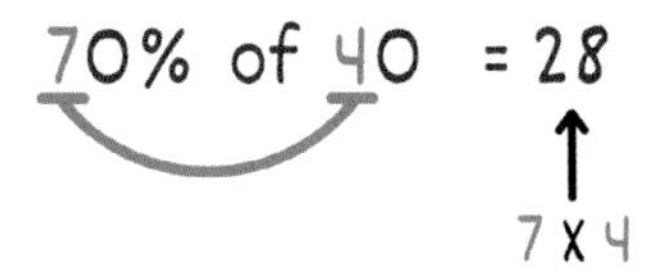

Let's keep the challenge streak going. What if we turned our two-digit number into a one-digit number like 40% of 9? Just like before, we multiply the first two digits (**4 x 9 = 36**) but this time make our answer 10 times smaller to get 3.6 (simply move the decimal point once to the left).

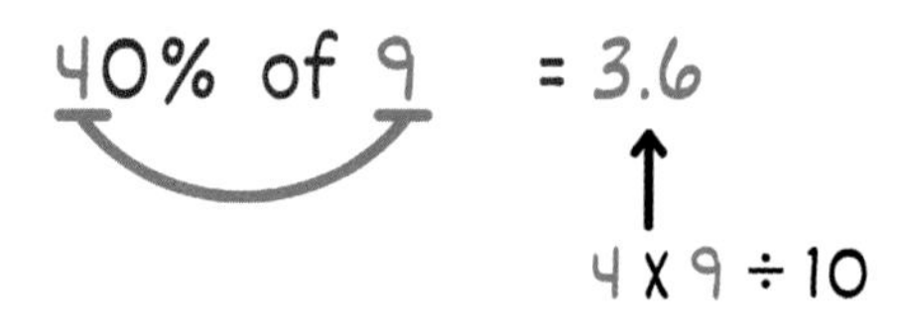

Practice

90% of 20

30% of 500

80% of 8

Why Does This Work?

Let's solve 30% of 80 by converting everything into fractions.

$$30\% \times 80 = \frac{30}{100} \times \frac{80}{1} = \frac{30 \times 80}{100 \times 1}$$

We can simplify this fraction by crossing out common factors in the numerator and denominator. Crossing off two zeros in the numerator and denominator, the fraction simplifies down to **3 x 8**, which equals to 24!

$$\frac{3\cancel{0} \times 8\cancel{0}}{1\cancel{0}\cancel{0} \times 1} = \frac{3 \times 8}{1 \times 1} = 3 \times 8 = 24$$

How Much Are You Saving with Discounts?

Let's say you are buying handmade beaded bracelets at a farmer's market.

If the bracelets are on sale at 30% off, how much will you save on $80 worth of bracelets? You can easily calculate 30% of 80 by multiplying the first two digits to get **3 x 8 = 24**. This means that you'll save $24 and only pay $56 for the bracelets!

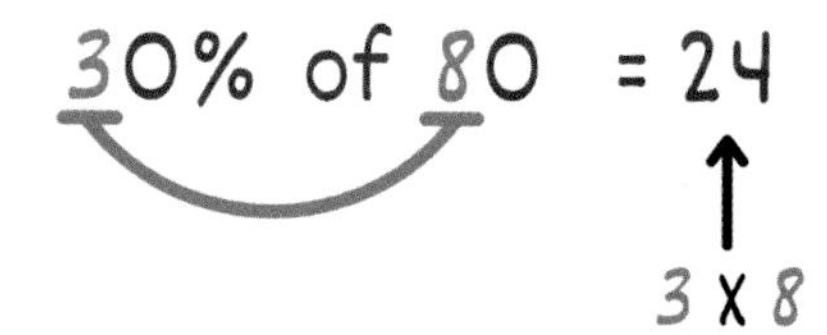

Here's a challenge: How much will you save on $800 worth of bracelets at 30% off? This time, you would still multiply the first two digits (**3 x 8 = 24**) but now add an extra 0 at the end to get 240. You'll be saving $240! And all we did was make our answer 10 times bigger.

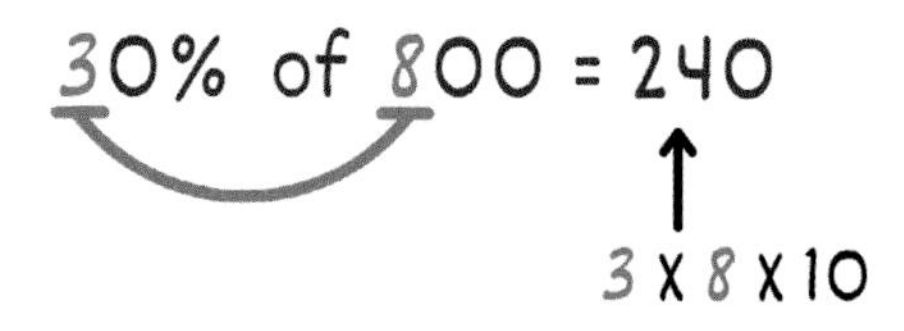

Chop Your Percentages

Who doesn't love a good percentage problem? You know, the ones where you're just taking 10% or 50% of something? Well, guess what, every percentage problem can be just as easy. All you must do is break your percentages down into easier ones like 50%, 10%, 5% or 1%. Let's give this a try and solve **26% of 80**.

26% of 80

Steps

1. Break your percentage down into a combination of 50%, 10%, 5% and 1%. Because **26 = 10 + 10 + 5 + 1**, we can break 26% down into 10% 10%, 5% and 1%.

26% of 80

10%

10%

5%

1%

Solve each smaller percentage starting with **10% of 80**. To find 10% of a number, simply move its decimal point once to the left.

26% of 80

10% ⟶ 8

10% ⟶ 8

5%

1%

3 Once you've solved **10% of 80**, you can easily solve **5% of 80**. Because 5 is half of 10, 5% of 80 must be half of 10% of 80. Half of 8 is 4!

26% of 80

10% → 8

10% → 8

5% → 4

1%

Finally, solve **1% of 80**. To do this, simply move the decimal point in 80 twice to the left.

26% of 80

10% → 8

10% → 8

5% → 4

1% → 0.8

5 Add all the percentage chunks you just calculated to get the final answer! So, **26% of 80 = 8 + 8 + 4 + 0.8 = 20.8**!

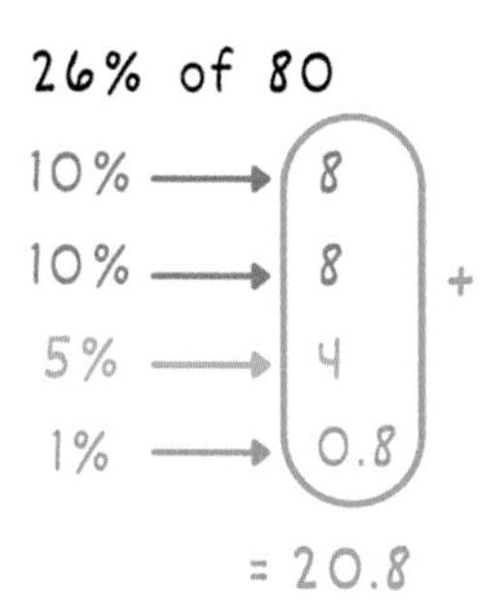

Practice

25% of 48

31% of 60

19% of 50 (challenge: try solving this using 19% = 20% - 1%)

Why Does This Work?

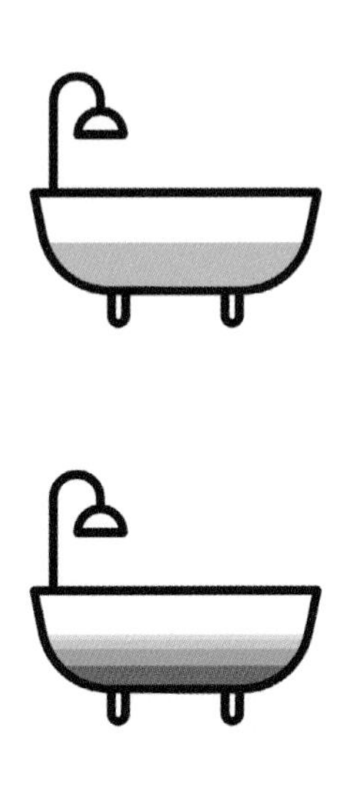

Think of it this way: A typical bathtub holds 80 gallons (303 L) of water. If you fill it to 26% of its capacity, you would have used 20.8 gallons (79 L) of water.

But you can achieve a 26% fill level for the bathtub without having to do it all at once. Rather, you can break it down into smaller fills, starting with filling 10% of the tub (which equals 8 gallons [30.3 L] of water), then another 10%, then 5% (4 gallons [15.1 L]) and finally the remaining 1% (0.8 gallon [3 L]). With these little splashes, you'll soon find yourself with a nice 26% filled bathtub!

How to Calculate Your Investment Gains!

This skill will come in handy in all sorts of situations. Take investing in the stock market, for instance. Let's say you take the plunge and buy $500 worth of Microsoft® shares. After a year, the stock increased by 13%. Not bad! So how much did your portfolio increase by?

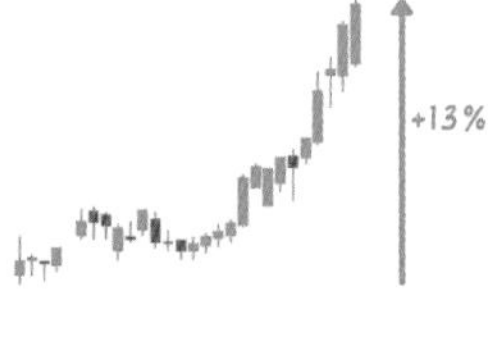

To figure out how much your Microsoft shares increased, all you need to do is find 13% of $500. This is where chopping your percentages comes in handy. Simply chop 13% into 10%, 1%, 1% and 1% and then add the answers! Your shares increased by $65, bringing your total portfolio value to $565!

13% of $500

10% → $50
1% → $5 +
1% → $5
1% → $5

= $65

Of-Is-What Magic

Have you come across word percentage problems? You know, the ones like:

What is 25% of 80?

15% of what number is 6?

What percent of 50 is 12?

Translating these into math problems can be a nightmare, but there is a trick you can use to make your life easier. Keep an eye out for the magic words *of*, *is* and *what*. With this handy toolbox, you'll never be confused again!

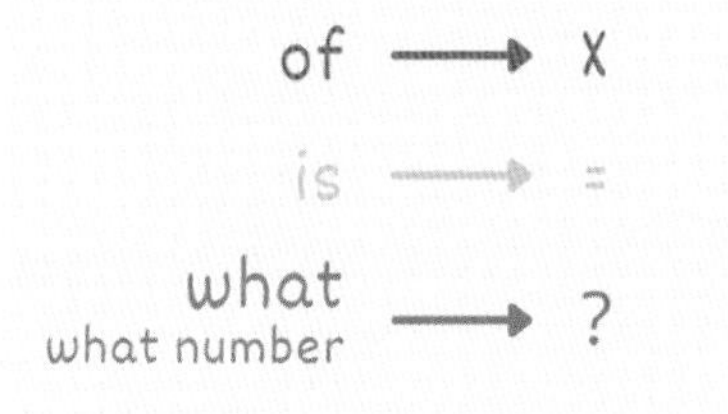

Steps

Here's how to use our handy toolbox to solve the first example equation "What is **25% of 80**?"

Replace "What" with a question mark (?).

What is **25% of 80?**
↓
?

2. Replace "is" with an equal sign (=).

What is 25% of 80?
↓ ↓
? =

Copy down 25% (because it's a number).

What is 25% of 80?
↓ ↓ ↓
? = 25%

Replace "of" with a multiplication sign (×).

What is 25% of 80?
↓ ↓ ↓
? = 25% X

5. Copy down 80 (because it's a number).

What is 25% of 80?
↓ ↓ ↓
? = 25% X 80

With these steps, you've translated "What is 25% of 80?" to "? = 25% × 80." Now you can solve for the question mark to find the answer!

$$? = 25\% \times 80$$
$$= 0.25 \times 80$$
$$= 20$$

Here's how you can use the toolbox to translate the next two examples:

What percent of 50 is 12?
↓ ↓ ↓
? X 50 = 12
? X 50 = 12

15% of what number is 6?
↓ ↓ ↓
15% X ? = 6
15% X ? = 6

Practice

What is 40% of 75?

20 is what percentage of 50?

70% of 20 is what number?

Why Does This Work?

It's straightforward why we replace "what" with a question mark, but why does "of" mean to multiply and "is" mean equal to?

Let's start off with "of." In the English language, "of" can be used in various contexts and one of its uses is to signify multiplication. It's possible that you are unconsciously using "of" to represent multiplication in your daily conversations! Let's look at a few examples. Suppose you're carrying 3 bags, each containing 2 water bottles. How many water bottles are you carrying in total? You can represent this as 3 bags "of" 2 water bottles, which translates to **3 x 2**, or 6 total bottles. Other examples include 8 pairs of socks, 2 boxes of 12 eggs and 7 groups of 10 students. There are many other ways to express multiplication, but "of" is frequently used when talking about percentages.

How about "is"? Well, "is" essentially means "equal to." This is particularly true in math problems where "is" is often a shorthand for "is equal to." For instance, the problem "**5% of 12** is what number?" is the equivalent to "**5% of 12** *is* equal to what number?" In math, "is" typically signifies equality. For example, "3 plus 5 is 8" implies **3 + 5 = 8**. The word "is" emphasizes that the left side of the equation is equal to the right.

How to Calculate Your Company's Worth with Fresh Funds

Let's say you start an online clothing brand that sells shirts made from sustainable bamboo fabric. You're growing rapidly and want to open your first physical shop in Los Angeles but need some extra money to do so. An investor decides to give you $50,000 in exchange for owning 20% of your company.

If $50,000 is worth 20% of your company, what is your company worth? Let's translate this sentence into an equation—$50,000 is 20% of what worth?

$$\begin{array}{ccccc} \$50{,}000 & \text{is} & 20\% & \text{of} & \text{what amount} \\ & \downarrow & & \downarrow & \downarrow \\ \$50{,}000 & = & 20\% & \times & ? \end{array}$$

Solving for the question mark, your company is worth $250,000. Nice!

$$\$50{,}000 = 20\% \times ?$$

$$\begin{array}{ccc} \$50{,}000 & = & 0.2 \times ? \\ \div 0.2 & & \div 0.2 \end{array}$$

$$\$250{,}000 = ?$$

Squares, Cubes and Roots Tricks

Feeling up for a little mental math challenge? Without using a calculator, can you tell me the values of these squares and square roots?

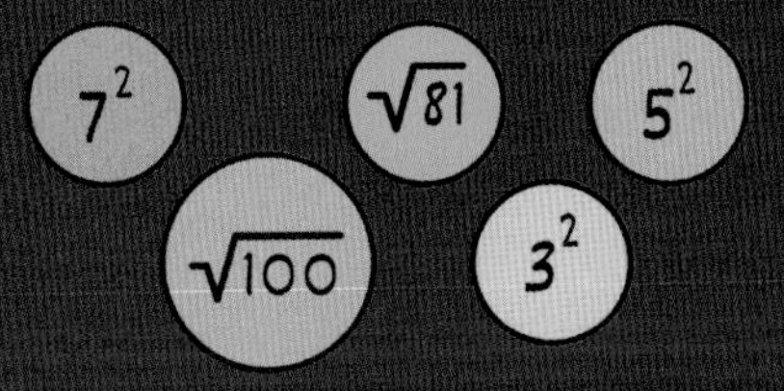

With those 2 to 10 times tables memorized, this should be a piece of cake. But what about bigger numbers or cubes and cube roots like the ones below? And when I say *cube*, I mean a number to the power of 3. On the other hand, the cube root of a number is a value that we can multiply 3 times to get that number.

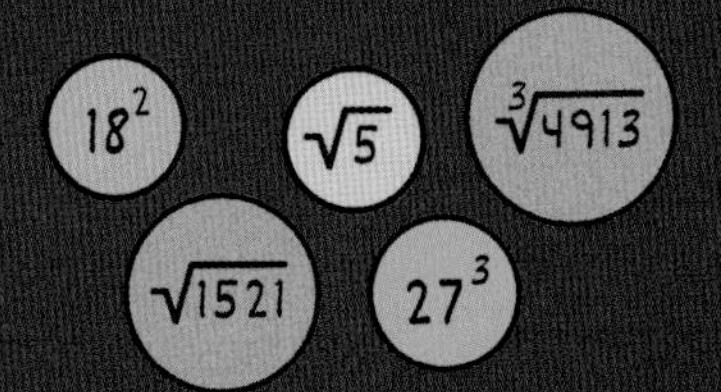

Don't let these numbers scare you. They may look impossible to solve, but by the end of this section, you'll be spitting out the answers to these without even thinking twice!

Need more convincing? Here's a little teaser to get you started—did you know that when you square or cube a number, the last digit will always be the same as the last digit of the original number squared or cubed? It's a super easy way to double-check your answer and make sure you're on the right track. Want to know the math behind it? Keep reading this chapter to find out!

$$73^2 = 5329 \quad (3^2 = 9) \qquad 12^3 = 1728 \quad (2^3 = 8)$$

By now you know that multiplication is just repeated addition and division is a game of repeated subtraction until zero. But what about an exponent? It's just repeated multiplication! Take 2^3 as an example—this is just 2 multiplied 3 times!

$$2^3 = 2 \times 2 \times 2 = 8$$

Exponents make things grow superfast, following the shape of a hockey stick! They're responsible for things like growing your investments, things that go viral on the internet and the spread of diseases.

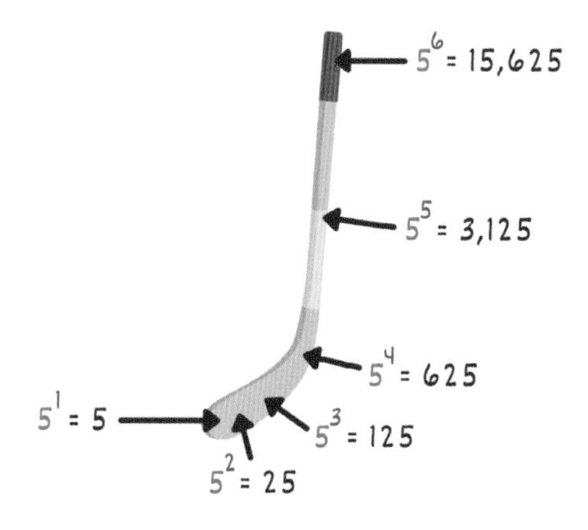

For example, let's say you post a video on YouTube that gets 2 views. If those 2 people forward the video to another 2 people and then those people each forward to another 2 people, your video will soon begin to go viral!

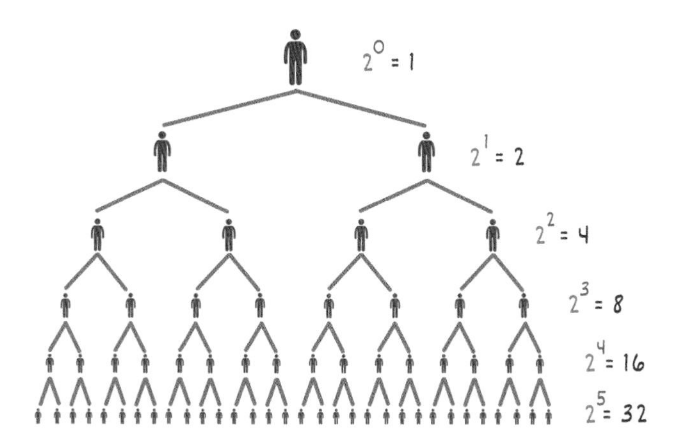

And now for a little terminology. The 2 in 2^3 is called the *base*, the 3 is called the *exponent* and the whole thing is called the *power*. We mostly deal with exponents that are 2 and 3, so we gave them special names—*squaring* and *cubing*.

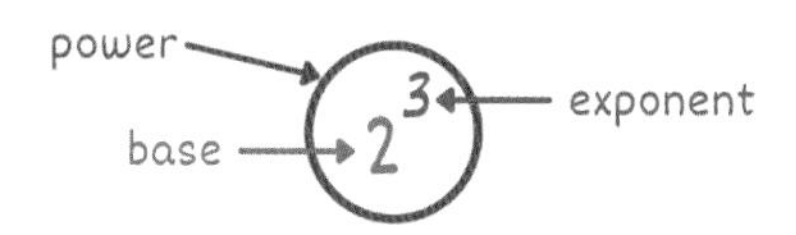

Want to go in reverse and find the base of a number squared or cubed? Just take the square root or cube root of that number.

$$5^2 = 5 \times 5 = 25$$

$$\downarrow$$

$$\sqrt{25} = \sqrt{5 \times 5} = 5$$

$$10^2 = 10 \times 10 = 100$$

$$\downarrow$$

$$\sqrt{100} = \sqrt{10 \times 10} = 10$$

All right, now that you're all caught up on the basics, let's have some fun and dive into the good stuff. Here we go!

Square Numbers From 1 to 99 in Your Head

Have you ever stepped inside an apartment in the hustle and bustle of New York City? If so, you know what I mean when I say they're tiny!

A typical New York City studio apartment measures anywhere from 20 to 25 feet (6.1 to 7.6 m) in both length and width. And with the cost of living there on the rise, the rental price has also gone through the roof—you're looking at around $5 per square foot.

So, if you're thinking of renting a cozy studio that's 21 by 21 feet (6.4 by 6.4 m), what's a good cost estimate to set aside for rent each month?

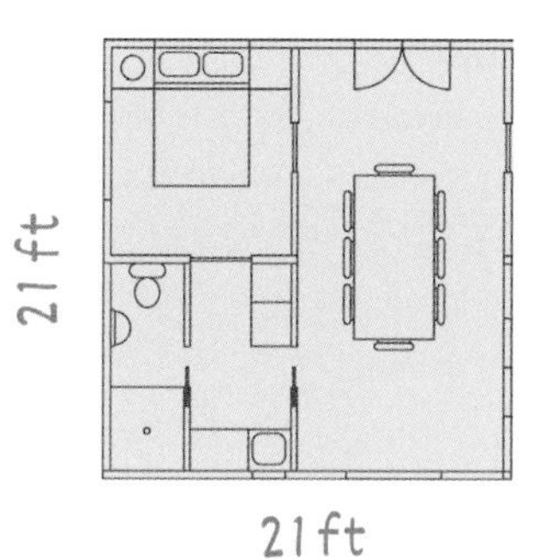

To solve this, let's first calculate the square footage (21^2) and then multiply it by $5.

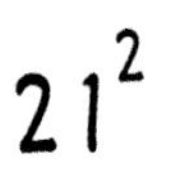

Steps

Divide your answer into three sections: first, middle and last.

$$21^2 = \underset{\text{first}}{\_\_}\ \underset{\text{middle}}{\_\_}\ \underset{\text{last}}{\_\_}$$

 Square the tens digit (2) and place it in the first section.

$$21^2 = 4\_\_\ \_\_$$

$$\uparrow 2^2$$

 Square the ones digit (1) and place it in the last section.

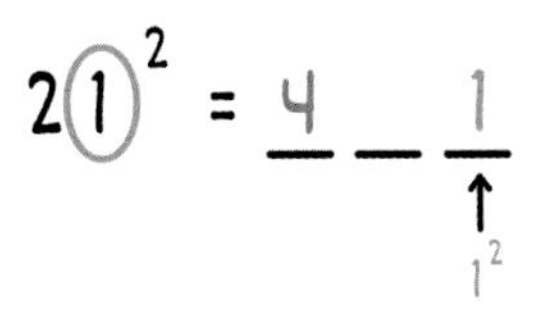

 Multiply the tens digit and the ones digit together and then multiply by 2 (**2 x 2 x 1 = 4**). Put this result in the middle section.

$$21^2 = 4\ 4\ 1$$

$$\uparrow 2(2 \times 1)$$

5 Check if the last section or the middle section has a two-digit number. If there is, carry the tens digit over. This example only has single digits in both of those sections, so you do not need to carry any numbers and **21^2 = 441**!

$$21^2 = 441$$

So how much would a 21 by 21–foot studio apartment in New York City roughly cost you? We just found that the apartment's area is 441 ft^2 (40.96 m^2), and because rent is about $5 per square foot, you're looking at a **441 x $5 = $2,205** monthly rental. That's a steep price for a small space, isn't it?

Up for another example? Let's tackle a problem that involves a bit of carrying. You'll need to do this whenever the middle or last sections have two-digit numbers inside of them.

$$43^2$$

Steps

 Just like before, divide your answer into three sections.

$$43^2 = \underset{\text{first}}{\_\_}\ \underset{\text{middle}}{\_\_}\ \underset{\text{last}}{\_\_}$$

 Square the tens digit (4) and place it in the answer's first section.

$$43^2 = \underset{4^2}{\underline{16}}\ \_\_\ \_\_$$

 Square the ones digit (3) and place it in the answer's last section.

$$43^2 = \underline{16}\ \_\_\ \underset{3^2}{\underline{9}}$$

4. Multiply the tens digit and the ones digit together and then multiply by 2 (**2 x 4 x 3 = 24**). Place this into the middle section of your answer.

$$43^2 = \underline{16}\ \underset{2(4 \times 3)}{\underline{24}}\ \underline{9}$$

5. Let's see if the last section has a two-digit number. Nope! How about the middle section? Yes! The middle section has a two-digit number (24), so you'll keep the ones digit (4) and carry the tens digit (2) and add it to the next section on the left (**16 + 2 = 18**). You'll get a final answer of 1,849. Don't worry if your first section becomes a two-digit number; it happens all the time!

$$43^2 = \overset{+2}{\underline{16}}\ \underline{4}\ \underline{9} = 1849$$

Practice

13^2 32^2 72^2

Why Does This Work?

Let's solve 21^2 using a bit of algebra. We can represent 21 as **$10a + b$**, where "a" is the tens digit and "b" is the ones digit (a = 2 and b = 1). Now, if we square our number, it becomes **$(10a + b)^2 = 100a^2 + 20ab + b^2$**. These three terms in the equation represent each section we solve for—the first section is $100a^2$, the middle section is 20ab and the last section is b^2. The 100 and 10 multiplied in the first and middle sections represent the hundreds and tens places!

We can also see this geometrically with a 21 × 21 square.

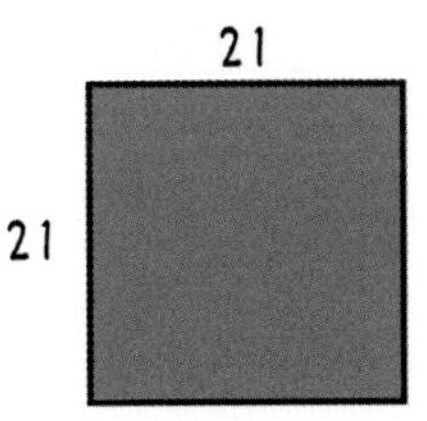

To get the area of the square, we can section the square into mini rectangles (**21 = 20 + 1**). Then, calculate the area of each mini rectangle and add all their areas up to get the total area. It's like a puzzle! The mini rectangles add up to **$20^2 + 2(20 \times 1) + 1^2$**, which follows the equation from earlier **$100a^2 + 20ab + b^2$**!

$21^2 = 400 + 20 + 20 + 1 = 441$

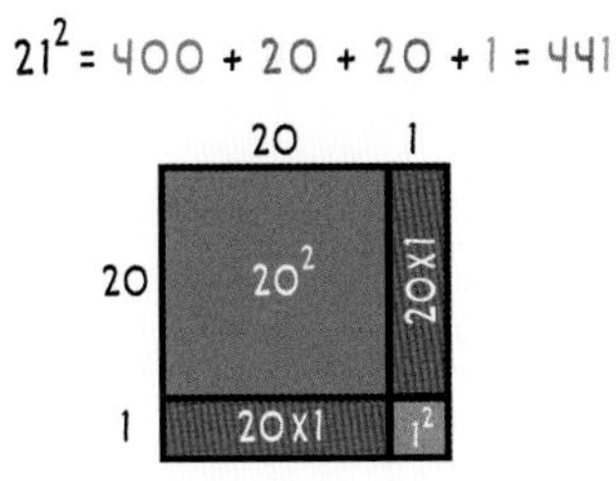

By using a bit of algebra and geometry, we can find the square of two-digit numbers in a quick and easy way. Give it a try with different numbers and see how it works for yourself.

Become a Squaring Master (100 to 999)

You've mastered the art of squaring numbers from 1 to 99 in your head, but are you ready to take it to the next level and square numbers from 100 to 1,000 like a math ninja? We'll be using the same logic as the last trick (page 140) but with a twist.

$$504^2$$

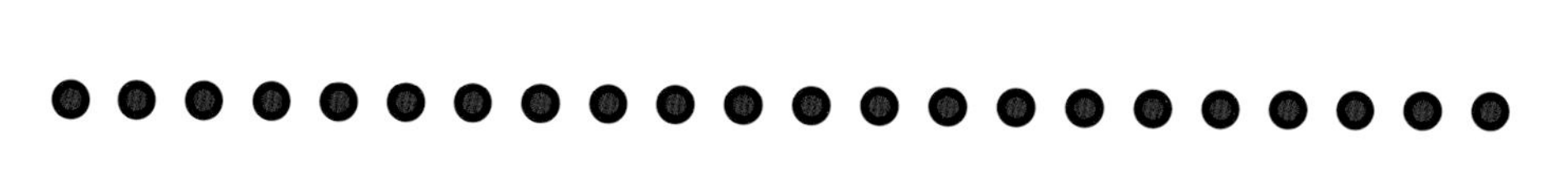

Steps

 Divide your answer into three sections: first, middle and last.

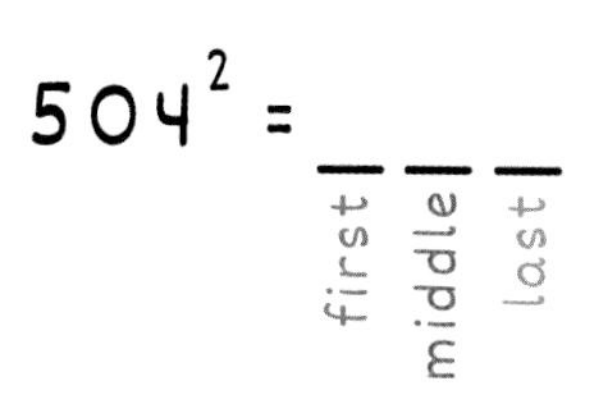

 Square the hundreds digit (5) and place it in the first section.

$$504^2 = \underline{25}\ \underline{\quad}\ \underline{\quad}$$

↑ 5^2

(3) Combine the tens digit and the ones digit (the 04 in 504). Then square it (**$4^2 = 16$**) and put the result into the last section.

$$504^2 = \underline{25}\ \underline{\quad}\ \underline{16}$$

↑ 4^2

Multiply the hundreds digit (5) with the combined tens and ones digits (04) and then multiply by 2 (**2 x 5 x 4 = 40**). Place this into the middle section of your answer.

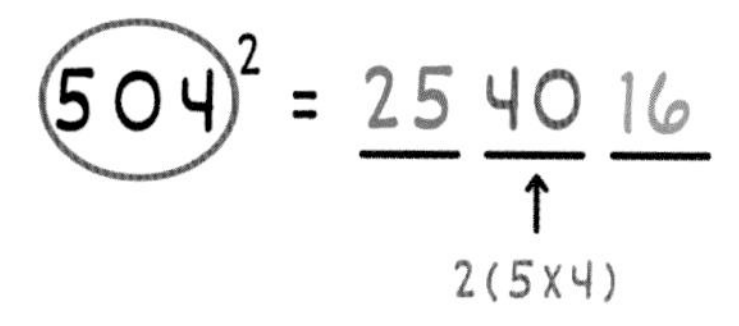

5. Check if the last section or the middle section has a three-digit number inside. If there is, carry the hundreds digit over to the next section on the left. In this case, because there are only two digits in both of those sections, there will be no carrying and you are done! In this example, $\mathbf{504^2 = 254{,}016}$!

$$504^2 = 254016$$

Can you believe we calculated such a big number like 254,016 in our heads? Imagine the reaction from your friends when you do that in front of them—it'll be priceless! But before you start practicing, let's try one more example where carrying is involved.

$$312^2$$

Steps

Divide your answer into three sections: first, middle and last.

$$312^2 = \underline{\quad}\ \underline{\quad}\ \underline{\quad}$$

first middle last

(2) Square the hundreds digit (3) and place it in the first section (**$3^2 = 9$**).

$$312^2 = \underset{\uparrow \atop 3^2}{\underline{9}}\ \underline{\quad}\ \underline{\quad}$$

 Combine the tens digit and the ones digit (the 12 in 312). Then square it (**$12^2 = 144$**) and put the result into the last section.

$$312^2 = \underline{9}\ \underline{\quad}\ \underset{\uparrow \atop 12^2}{\underline{144}}$$

(4) Multiply the hundreds digit (3) with the combined tens and ones digit (12) and then multiply by 2 (**$2 \times 3 \times 12 = 72$**). Place this into the middle section of your answer.

$$312^2 = \underline{9}\ \underset{\uparrow \atop 2(3 \times 12)}{\underline{72}}\ \underline{144}$$

(5) Let's see if the last section has a three-digit number. And it does (144)! So you'll keep the ones and tens digit (44), carry the hundreds digit (1) and add it to the next section on the left (**$72 + 1 = 73$**). Does the middle section have a three-digit number? Not this time! So you're all done and the final answer is 97,344. Just remember, the first section is allowed to have three digits, so don't worry if yours does!

$$312^2 = \underline{9}\ \overset{+1}{\underline{72}}\ \underline{44} = 97344$$

Practice

111^2 $\quad$ 132^2 $\quad$ 541^2

Why Does This Work?

When we algebraically squared two-digit numbers in the previous example we represented our two-digit numbers as 10a+b, where "a" is the tens digit and "b" is the ones digit (for example, for 21, a = 2 and b = 1). Similarly, when we square three-digit numbers we can represent our numbers as 100a+b, where "a" is the hundreds digit and "b" is the tens and ones digit combined (for example, for 312, a = 3 and b = 12). This expands out to $\mathbf{10{,}000a^2 + 200ab + b^2}$ where the first section is $10{,}000a^2$, the middle section is 200ab and the last section is b^2.

We can also see this geometrically with a 312 × 312 square.

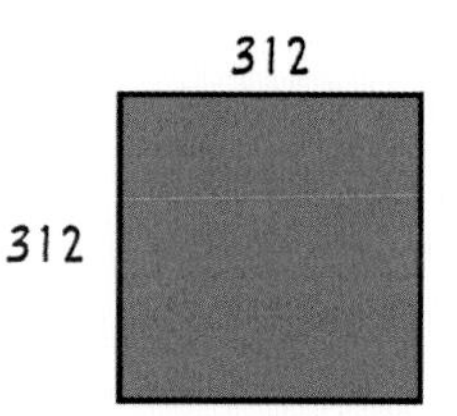

To get the area of the square, we can section the square into mini rectangles (**312 = 300 + 12**). Then, we calculate the area of each mini rectangle and add all their areas up for the total area. The mini rectangles add up to $\mathbf{300^2 + 2(300 \times 12) + 12^2}$, which follows the equation from earlier: $\mathbf{10{,}000a^2 + 200ab + b^2}$!

$$312^2 = 90000 + 3600 + 3600 + 144 = 97344$$

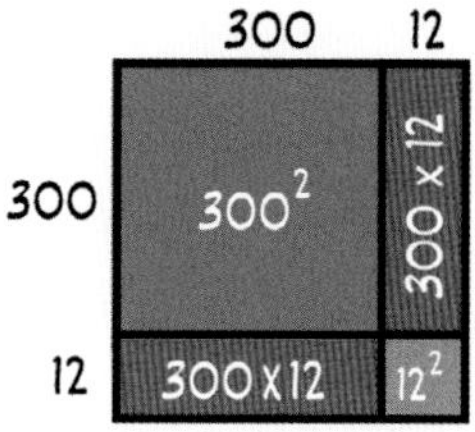

15^2, 25^2, 35^2, 45^2, 55^2, 65^2, 75^2, 85^2, 95^2 in 3 Seconds!

When I say you can square two-digit numbers that end in a 5 in three seconds, I do mean three seconds flat. Are you curious to know how? Here's the secret!

$$35^2$$

Steps

First, multiply the tens digit (3) by one more than itself (**3 x [3 + 1] = 3 x 4 = 12**).

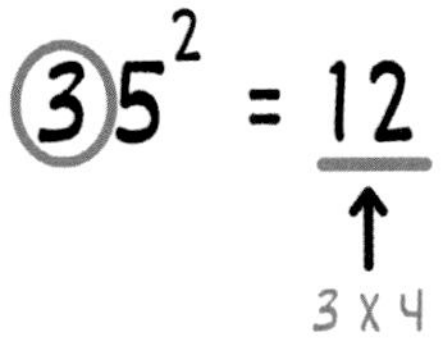

Append a 25 at the end, and that is your answer!

$$35^2 = 1225$$

Easy, don't you think? Now how quickly can you get 75^2? Multiplying 7 by one more than itself (**7 x 8 = 56**) and then adding 25 at the end, we get 5,625. Remember to always put a 25 at the end of your answer, like a cherry on top!

Practice

Why Does This Work?

Let's solve 35^2 algebraically. We can represent 35 as 10a+b, where "a" is the tens digit and "b" is the ones digit (a = 3 and b = 5). When we square it, we get $(10a+b)^2 = 100a^2 + 20ab + b^2$. Because b is always 5, we can simplify it to $100a^2 + 100a + 25$. And if you look closely, you can see that the hundreds place is a(a+1), which is the tens digit multiplied by one more than itself, and the ones digit will always be 25! Isn't that cool?

We can also see this geometrically with a 35 × 35 square.

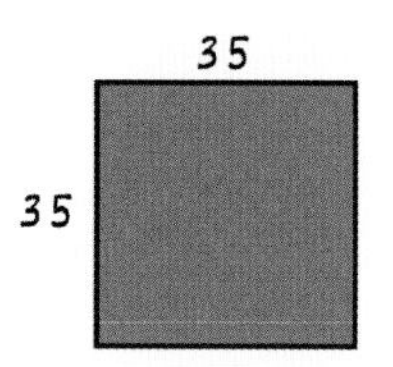

To get the area of the square, we can imagine breaking it into mini rectangles (split by the tens digit and the ones digit), calculating the area of each mini rectangle and adding them up to get the total area.

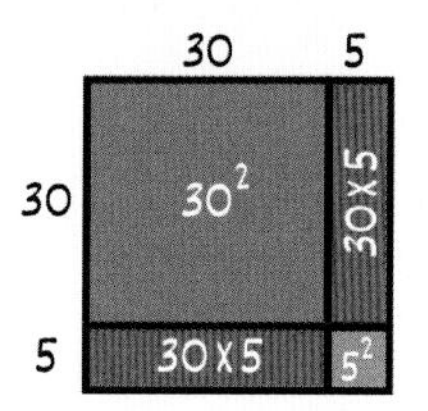

All we're doing in this trick is just moving around some mini rectangles to create a new rectangle (**30 x 40 = 1,200**) and an extra little square (**5 x 5 = 25**)! It's like a game of Tetris®!

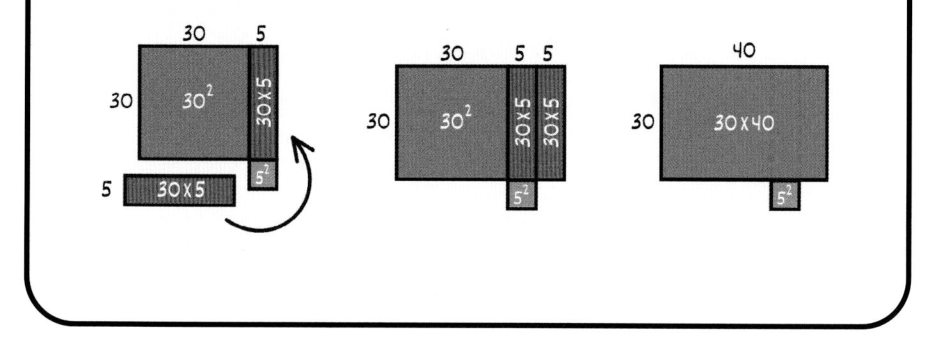

The Trick to Cubing Numbers

Congratulations, my friend! You have made it to the chapter that's the longest, most challenging, but most rewarding of all. Take a moment to appreciate how far you've come! If you're up for tackling a big challenge today, come with me as I reveal a trick that will teach you how to cube a number in your head by using a grid.

$$12^3$$

How would you take 12 and cube it in your head? To do this, you'll first need to know the cubes for numbers 1 to 9, but trust me, it'll be worth it!

$$1^3 = 1$$

$$2^3 = 8$$

$$3^3 = 27$$

$$4^3 = 64$$

$$5^3 = 125$$

$$6^3 = 216$$

$$7^3 = 343$$

$$8^3 = 512$$

$$9^3 = 729$$

Feel free to reference this list while we work through the steps to solve 12^3. With a bit of practice, you'll have this list remembered in no time!

Steps

1. Let's represent the tens digit of our number as "a" and the ones digit as "b." Combining the digits together, our number will be represented as "ab." For the number in our example, 12, a = 1 and b = 2.

2. This is where things get interesting. Draw a grid and write a^3, a^2b, $2a^2b$, ab^2, $2ab^2$ and b^3 inside. This grid will help us easily break down the multiplication process so that we can do it in our heads with some practice. Do you notice any patterns? The terms in the second row are just double the terms above them in the first row! Keep this in mind, as it will come in handy later on.

a^3	a^2b	ab^2	b^3
	$2a^2b$	$2ab^2$	

3. Because a = 1 in 12, we can plug 1 into each "a" and our grid will simplify down significantly. That's right—cubing numbers that either start or end with 1 will be much easier to solve! So if you ever come across a number like 15, 31 or 91, you can use a simplified grid instead.

1	b	b^2	b^3
	$2b$	$2b^2$	

Great! Because b = 2 in 12, we can plug 2 into each "b." I suggest doing this from left to right starting with the first row.

1	2	b^2	b^3
	2b	$2b^2$	

Next, plug 2 into b^2 in the third column. Plugging in 2 we get **2^2 = 2 x 2 = 4**.

1	2	4	b^3
	2b	$2b^2$	

6

Finally let's plug 2 into b^3 in the fourth column. This becomes **2^3 = 2 x 2 x 2 = 8**.

1	2	4	8
	2b	$2b^2$	

Now for the second row. You can plug 2 into each "b" again, but there is an even faster way to do this. What did we establish about the second row in step 2? We established that it's double the value of the first row! So instead of plugging 2 into each "b," we can simply double the numbers in the first row. Let's first double the 2 in the second column to get 4.

1	2x2	4	8
	↓4	$2b^2$	

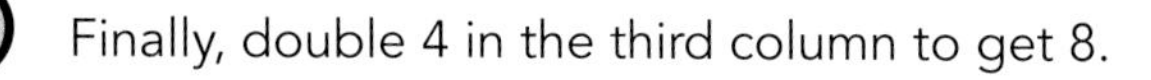

(8) Finally, double 4 in the third column to get 8.

1	2	2×4	8
	4	↓8	

(9) You've done an excellent job filling in the grid so far, and you're almost done! All you have left is to add all the numbers up to get the final answer. Treat the top row as a number (1,248) and the bottom row as another number (480) and add them as you traditionally would from right to left. Just like when you add two numbers, if any column (except the left-most one) adds up to be 10 or above, simply keep the ones digit and carry the tens digit over to the second column on the left! In the third column, **4 + 8 = 12**, so we will keep 2 and carry 1 over to the second column.

		+1		
	1	2	4	8
+		4	8	
	1	7	2	8

Combine all the digits you just added to get the final answer of **12^3 = 1,728**!

$$12^3 = 1728$$

Not bad at all, right? Now let's try a slightly more challenging one that will involve a bit of carrying when filling out your grid. This one will be a little bit longer, but don't worry, I'll be here to guide you every step of the way. Remember, the key is to take it one step at a time and not to get overwhelmed!

$$23^3$$

Steps

1. Draw out your grid where a = 2 and b = 3.

a^3	a^2b	ab^2	b^3
	$2a^2b$	$2ab^2$	

Just like before, let's plug a = 2 and b = 3 in our grid from left to right starting with a^3 in the first column. This becomes **$2^3 = 2 \times 2 \times 2 = 8$**.

8	a^2b	ab^2	b^3
	$2a^2b$	$2ab^2$	

Next up we have **$a^2b = 2^2 \times 3 = 12$**. Because 12 is a two-digit number and each column can only have one digit, you'll need to keep the ones digit (2) and carry the tens digit (1) over to the column on the left.

+1 8	2	ab^2	b^3
	$2a^2b$	$2ab^2$	

Next we have **$ab^2 = 2 \times 3^2 = 18$**. Again, because 18 is a two-digit number, keep the ones digit (8) and carry the tens digit (1) over to the next column.

+1 8	+1 2	8	b^3
	$2a^2b$	$2ab^2$	

The last column in the first row is $b^3 = 3^3 = 3 \times 3 \times 3 = 27$. Keep the ones digit (7) and carry the tens digit (2) over to the next column.

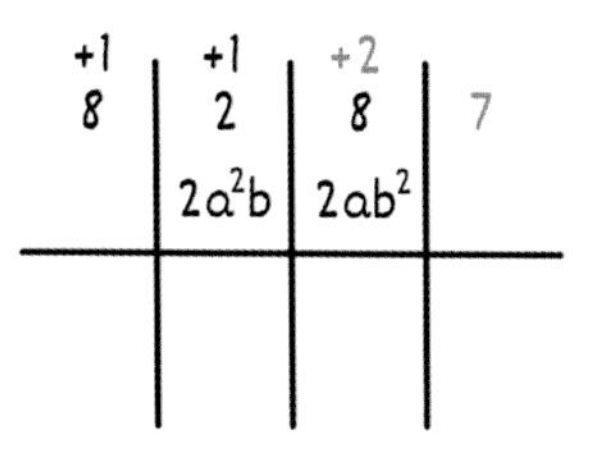

6

Time to fill out the bottom row! For the second column, double 12 to get 24 (alternatively, you could calculate the full thing out $2a^2b = 2 \times 2^2 \times 3 = 24$). Keep the 4 and carry the 2 over to the next column on the left.

7

Lastly, double 18 to get 36 (or calculate it out $2ab^2 = 2 \times 2 \times 3^2 = 36$). Keep the 6 and carry the 3 over to the next column on the left.

8 Woohoo! You've filled out the entire grid! Now for the easy part. Let's add all the numbers up in each column. In the second and third columns from right to left, our numbers add up to two digits (16 and 11), so make sure you keep the ones digit and carry the tens digit over to the next column on the left while adding!

	+1	+1		
	+2	+3		
	+1	+1	+2	
	8	2	8	7
+		4	6	
	12	1	6	7

Adding everything up we get our final answer of **23^3 = 12,167**. You did it!

$$23^3 = 12167$$

Practice

11^3 17^3 32^3

Why Does This Work?

When we square a number, we create a square, but when we cube a number, what do you think we create? A cube!

$(ab)^2$

ab

ab

$(ab)^3$

ab ab

ab

Now how do we find the volume of a cube? One way is by cutting it up into smaller chunks, finding the volume of each chunk and then adding them all together. But how do you know where to slice the cube? There are many ways, but an easy way is by breaking each side down by place values. Let's represent our number (12) algebraically as (10a+b) where "a" is the tens digit (1) and "b" is the ones digit (2). When we cube this and multiply out the binomials, we get $\mathbf{(10a + b)^3 = 1{,}000a^3 + 300a^2b + 30ab^2 + b^3}$.

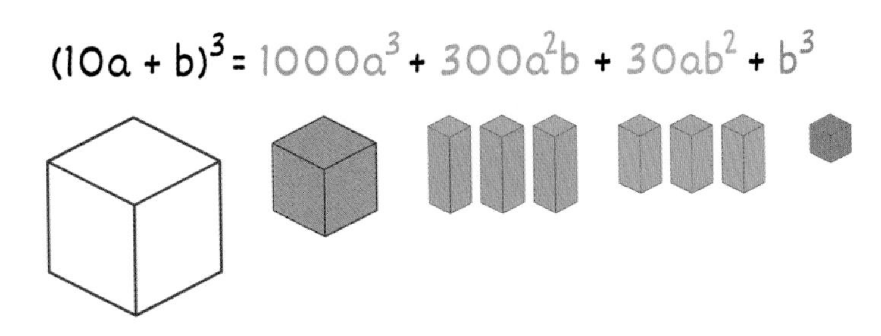

Wait, does something about this equation look familiar? We can break $300a^2b$ and $30ab^2$ down to get $\mathbf{1{,}000a^3 + 200a^2b + 100a^2b + 20ab^2 + 10ab^2 + b^3}$. These six terms are the same six terms we added together in our grid (the coefficient numbers in front of them determine their place value in the thousands, hundreds, tens and ones places)!

$$(10a + b)^3 = 1000a^3 + 200a^2b + 100a^2b + 20ab^2 + 10ab^2 + b^3$$

a^3	a^2b	ab^2	b^3
	$2a^2b$	$2ab^2$	

Why did we separate $300a^2b$ and $30ab^2$ into ($\mathbf{200a^2b + 100a^2b}$) and ($\mathbf{20ab^2 + 10ab^2}$)? Because it is much easier to mentally calculate a term like a^2b and then double it than it is to calculate the full $3a^2b$.So, let's get slicing and see what you come up with!

What Is Napkin Math?

Have you ever heard of *napkin math*? It's a colloquial term used in the finance industry that refers to quick calculations made during informal settings, such as business dinners. For example, if you are managing an investment portfolio and are considering investing around $100,000 in a young company expected to grow about 20% annually, you can use the compound interest formula to quickly estimate the value of your investment in 3 years.

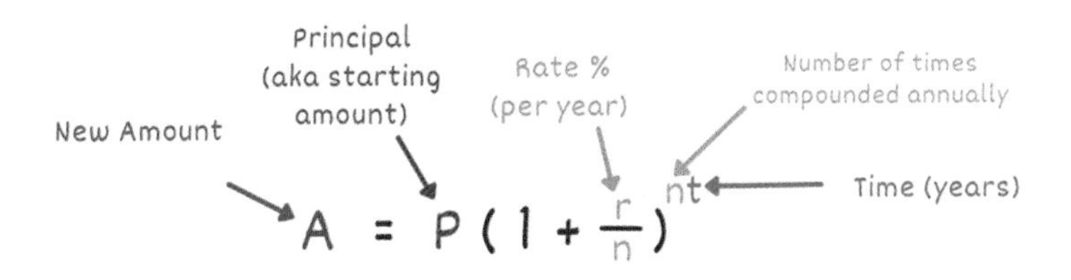

Let's input r = 20%, n = 1 (it grows 20% annually) and t = 3 (investment worth in 3 years) into the formula. These are easy-to-use, estimated numbers. Thus, napkin math!

$$A = 100{,}000 \times 1.2^3$$

Instead of solving 1.2^3, let's solve 12^3 and then divide by 1,000. This works because $\mathbf{12^3 = (10 \times 1.2)^3 = 10^3 \times 1.2^3}$. And how do you solve 12^3 in your head? We solved this in the first example in this chapter, but try it on your own this time and then double-check your answer with the provided example on page 153!

With that, you just completed the toughest trick of all in this entire book! Now that you've nailed this one, everything else will seem easy in comparison. Excellent work!

Become a Square Root Calculating Machine!

Tired of relying on your calculator for square roots? I'm sure you know perfect squares like 4, 9, 16 and 25 like the back of your hand, but what about those tricky nonperfect squares like 3, 27 or 105? Well, I've got a fun trick to help you estimate the square root of any number! Just a heads up, it does require you to know your perfect squares, but don't worry, I've got a quick refresher for you.

$1^2 = 1$

$2^2 = 4$

$3^2 = 9$

$4^2 = 16$

$5^2 = 25$

$6^2 = 36$

$7^2 = 49$

$8^2 = 64$

$9^2 = 81$

Using these perfect squares, let's find the square root of 27.

$$\sqrt{27}$$

Steps

1. First ask yourself, "What is the closest perfect square that is below 27?" That would be 25 from 5^2 (not 36 from 6^2 because we are looking for a perfect square below 27). Write $\sqrt{25}$ right next to $\sqrt{27}$.

2. Now let's find the whole number part of our answer by taking the square root of the perfect square you just wrote down (**$\sqrt{25} = 5$**).

3. The next part of our answer will always be a fraction. To find the numerator, subtract the perfect square you wrote down (25) from the original number (27) to get **27 - 25 = 2**.

4. To find the denominator (bottom), take your whole number part and double it (**5 x 2 = 10**).

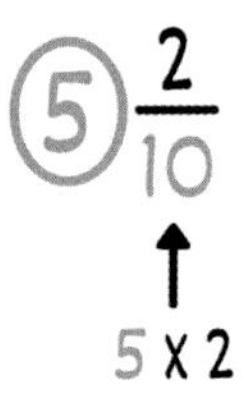

5. And that's it! You can also go ahead and simplify your fraction (or I usually like to turn my answer into a decimal).

$$5\frac{2}{10} = 5\frac{1}{5} = 5.2$$

You just mentally calculated the square root of 27 to be 5.2. Nice! However, keep in mind that this is an estimate. All square roots of nonperfect squares will always be irrational decimals that go on forever and forever. Let's double-check on a calculator and see **($\sqrt{27}$ = 5.196152422706631880582339...)**. So, 5.2 looks like a pretty good approximation!

Practice

$\sqrt{10}$ $\sqrt{52}$ $\sqrt{93}$

Why Does This Work?

We can represent any nonperfect square number by $N = a^2 + b$, where "a" is the root of the closest perfect square below N and "b" is the amount extra added to the perfect square to equal N. In our example, N = 27, a = 5 and b = 2. For another example, if N = 85, then a = 9 and b = 4. Now let's shuffle the equation and solve for b ($b = N - a^2$).

$$N = a^2 + b$$

$$b = N - a^2$$

Let's keep going with this and solve for $\sqrt{N}$. We'll be shifting and shuffling terms around, so keep your eyes peeled and try to follow along! We'll first factor $(N - a^2)$ into $(\sqrt{N} + a)(\sqrt{N} - a)$, then divide both sides by $(\sqrt{N} + a)$ and finally add "a" on both sides.

$$b = N - a^2$$

$$b = (\sqrt{N} + a)(\sqrt{N} - a)$$

$$(\sqrt{N} - a) = \frac{b}{(\sqrt{N} + a)}$$

$$\sqrt{N} = a + \frac{b}{(\sqrt{N} + a)}$$

(continued)

If you're wondering how I factored $(N - a^2)$ into $(\sqrt{N} + a)(\sqrt{N} - a)$, here's an explanation on how I did it. There are several ways you can go about factoring this and there is no set procedure to follow—it's all by trial and error! The first thing I observed was that a^2 has a negative sign in front of it. This means that one of our factors will be positive and the other will be negative **(__ + __)(__ - __)**. Next, I asked myself, "What two values (besides 1) multiply together to get N?" Well, $\sqrt{N} \times \sqrt{N} = N$, so we know that $(\sqrt{N} + \_\_)(\sqrt{N} - \_\_)$. Finally, I asked myself, "What two values (besides 1) multiply together to get a^2?" We know that $a \times a = a^2$, so this becomes $(\sqrt{N} + a)(\sqrt{N} - a)$! We can multiply all the terms in each binomial to check that we factored this correctly: $\sqrt{N}\sqrt{N} + a\sqrt{N} - a\sqrt{N} + (a)(-a) = N - a^2$.

All right, now back to seeing why this trick works! Take a close look at $(\sqrt{N} + a)$ at the bottom. Because $\sqrt{N}$ will always be an irrational number that is very close to value with "a," we can estimate $(\sqrt{N} + a) = 2a$.

$$\sqrt{N} = a + \frac{b}{(\sqrt{N} + a)}$$

$$\sqrt{N} = a + \frac{b}{2a}$$

All right, let's plug $(N - a^2)$ into "b" and take a step back. Does this equation look familiar? The three steps we used to estimate $\sqrt{27}$ follows this equation! If you plug in a = 5 and N = 27, you'll see that these are the same steps we took earlier. I know that was a lot of information to process, but kudos to you for sticking with it and following along with me! You're doing great!

$$\sqrt{N} = a + \frac{N - a^2}{2a}$$

$$\sqrt{N} = 5 + \frac{27 - 25}{2 \times 5} = 5\frac{2}{10} = 5\frac{1}{5} = \mathbf{5.2}$$

How Big Is Your Bathroom?

You're looking at a blueprint of your bathroom. If it says that your bathroom is 105 feet² (9.75 m²) and looks close to the shape of a square, would you be able to fit a bathtub along one of the walls if a bathtub is 5 feet (1.52 m) long?

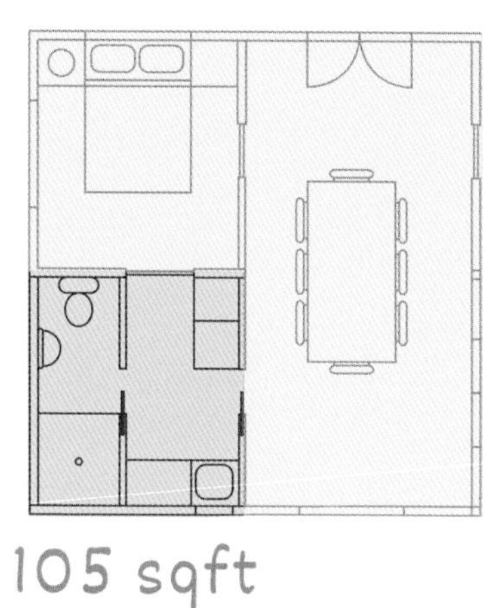

To solve this, let's find the square root of 105. The closest square below 105 is 100 (from 10^2). This means the whole number part will be 10. The numerator will be 5 (**105 - 100**), the denominator 20 (**10 x 2**) and our final answer is 10 and $\frac{5}{20}$ feet, or 10.25 feet (3.12 m). This means you will be able to fit a bathtub inside!

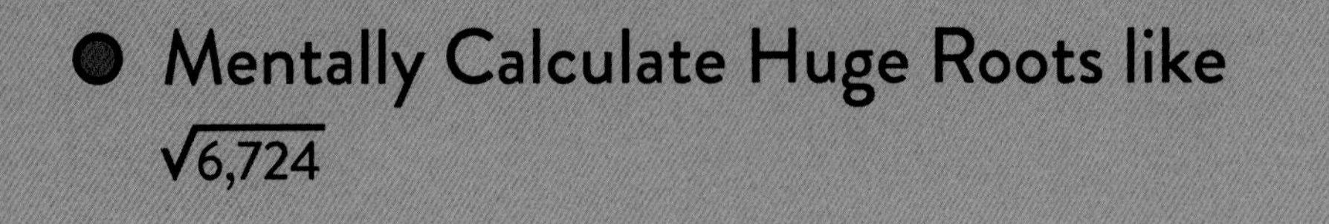

Mentally Calculate Huge Roots like $\sqrt{6,724}$

Want to calculate enormous roots like $\sqrt{529}$, $\sqrt{6,724}$ and $\sqrt{9,801}$ by just using your raw brainpower? Here's a mental math trick that will allow you to find the square root of any perfect square. All you need to know are your perfect squares from 1 to 9 and you're all set!

$1^2 = 1$

$2^2 = 4$

$3^2 = 9$

$4^2 = 16$

$5^2 = 25$

$6^2 = 36$

$7^2 = 49$

$8^2 = 64$

$9^2 = 81$

For this example, let's find the square root of 6,724.

$$\sqrt{6724}$$

Steps

1. First, split your number underneath the square root symbol and dissect it into two sections. The section on the right will include the last two digits, and the section on the left will include everything else. For example, 6,724 will split into 67|24, while a three-digit number like 529 will split into 5|29.

$$\underline{67}\,\underline{24}$$

2. Now, let's focus on the left section of our number (67). Which two squares is 67 nestled between? It's between $8^2 = 64$ and $9^2 = 81$. Just take the lower number of those (8 is lower out of 8 and 9) and that will be the first digit in your answer.

$6^2 = 36$

$7^2 = 49$

$8^2 = 64$ ← 67

$9^2 = 81$

6724 = 8

3. Let's shift our focus to the right section of our number (24) and take a closer look at the last digit (4). This is where things get interesting. Think back to all the squares you have memorized. Which perfect square ends with the same number as yours (4)? Both $\mathbf{2^2 = 4}$ and $\mathbf{8^2 = 64}$ end in a 4. One of these (2 or 8) will be the second digit in your answer, so write both down and we'll figure out which one to pick in the next step.

$1^2 = 1$

→ $2^2 = 4$

$3^2 = 9$

$4^2 = 16$

$5^2 = 25$

$6^2 = 36$

$7^2 = 49$

→ $8^2 = 64$

$9^2 = 81$

6724 = 8 _

↑

2 or 8

(4) It's time to pick the second digit of our answer (2 or 8). Let's take a quick look back at our first digit (8). To help us decide, we'll multiply the first digit with one more than itself (**8 x 9 = 72**) and compare 72 with the number in the first section (the 67 from 6,724). Is 67 bigger or smaller than 72? It's smaller! So, we'll pick the smaller of 2 and 8 for our answer.

$1^2 = 1$
$2^2 = 4$
$3^2 = 9$
$4^2 = 16$
$5^2 = 25$
$6^2 = 36$
$7^2 = 49$
$8^2 = 64$
$9^2 = 81$

$8 \times 9 = 72$
$67 < 72$

6724 = 82

2 or 8

And there you have it! By carefully choosing 2 out of 2 and 8, we arrive at the final answer of **$\sqrt{6,724} = 82$**. It's like solving a riddle, isn't it?

6724 = 82

Practice

$\sqrt{169}$

$\sqrt{1521}$

$\sqrt{9216}$

Why Does This Work?

Once you have 1^2, 2^2, 3^2, 4^2, 5^2, 6^2, 7^2, 8^2 and 9^2 memorized, you'll also know the squares for 10^2, 20^2, 30^2, 40^2, 50^2, 60^2, 70^2, 80^2 and 90^2. For instance, did you know that $2^2 = 4$ and $20^2 = 400$? Or that $5^2 = 25$ and $50^2 = 2{,}500$? And how about $8^2 = 64$ and $80^2 = 6{,}400$? See the pattern? The squares for two-digit numbers just have an extra 00 at the end because they're 100 times bigger!

Now let's see how we solved our example $6{,}724 = 82^2$. Would you agree with me that 82^2 is between 80^2 and 90^2? During the first step, we only focused on the 67 in 6724 and isolated it as 6700. Because 6700 is between 6400 (80^2) and 8100 (90^2), we know that the first digit must be 8 (the 8 in 82 represents 80 because it is in the tens digit).

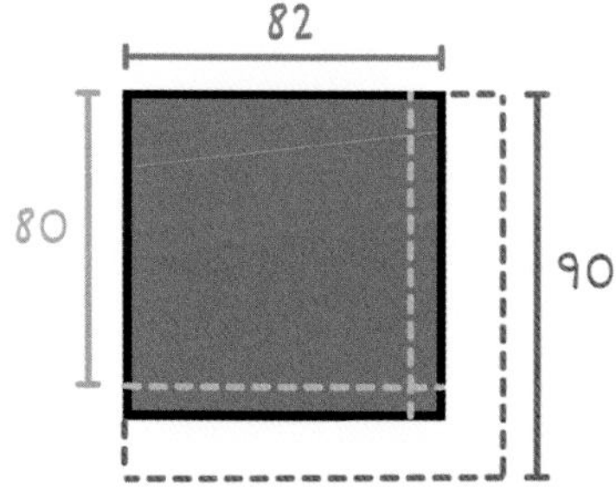

All right, let's keep going! Why did the ones digit in 6724 equal the ones digit in our squared number? Let's use a bit of algebra to make it simple. We can represent our two-digit number as $10a+b$, where "a" is the tens digit and "b" is the ones digit (a = 8 and b = 2 for 82). Now, if we square our number and multiply the binomials, we get $(10a + b)^2 = (10a + b)(10a + b) = 100a^2 + 20ab + b^2$. Do you see how all the terms except b^2 have a coefficient that is a power of 10 in front of it? This means that only b^2 will affect the ones digit in our number square! That's why the ones digit in 6724 must equal the ones digit in $2^2 = 4$. Interesting, isn't it?

How to Properly Buy a Computer Monitor

Have you ever purchased a computer monitor? If so, you may have noticed that the size of a monitor is determined by its diagonal length and not its height or width!

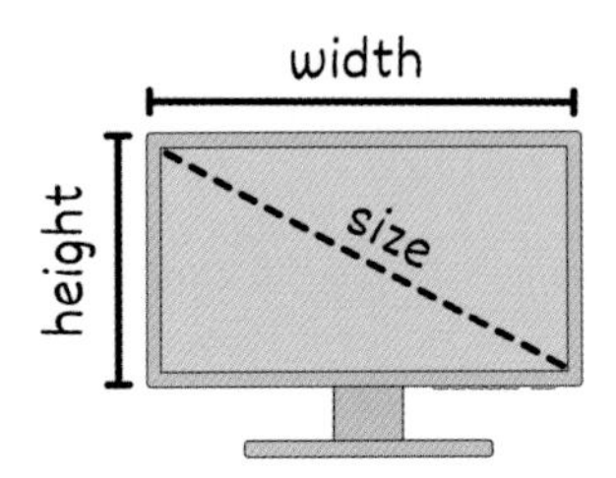

For example, if you have a computer monitor that measures 12.35 inches (31.4 cm) in height and 24 inches (61 cm) in width, you can calculate the diagonal length of the monitor by first using the *Pythagorean Theorem*. The theorem states that in a right triangle, the sum of the square of the two shorter sides (a and b) is equal to the square of the length of the longest side, also known as the *hypotenuse* (c).

a c b

$$c^2 = a^2 + b^2$$

$$c = \sqrt{a^2 + b^2}$$

If we plug the sides of the computer monitor as a = 24 inches and b = 12.35 inches, we can calculate the monitor's diagonal length as $c = \sqrt{(24^2 + 12.35^2)} \approx \sqrt{729}$.

Now comes the fun part. How can you solve $\sqrt{729}$ without using a calculator? Just like before, let's break 729 down into 7|29. Now, 7 is nestled between $2^2 = 4$ and $3^2 = 9$, so let's pick the lower number, or 2, for the first part of our answer. For the second part, let's focus on the last digit, 9. By looking at our perfect squares, we can see that both $3^2 = 9$ and $7^2 = 49$ end in a 9, so our answer's second digit will be either 3 or 7. To determine which one it is, let's go back to the answer's first digit, 2, and multiply it with one more than itself ($2 \times 3 = 6$). Is 7 bigger or smaller than 6? It's bigger! So, let's pick the bigger number out of 3 and 7, which is 7. We'll get a final answer of $\sqrt{729} = 27$. This means that the size of our computer monitor is 27 inches (68.6 cm)!

Cube Root Magic: From 1 to 999,999

Here's a fun math trick that will amaze your friends and family! Ask a friend to choose a whole number between 1 and 100 (and keep it a secret) and have them cube it on a calculator. Then, have them show you the new number. In no time, you'll be able to tell them their original number by quickly taking the cube root in your head. Are you ready to learn the secret and amaze your friends?

This trick does require you to remember the cubes of 1 to 9, but don't worry, it's not too difficult and it will be worth it in the end!

$1^3 = 1$

$2^3 = 8$

$3^3 = 27$

$4^3 = 64$

$5^3 = 125$

$6^3 = 216$

$7^3 = 343$

$8^3 = 512$

$9^3 = 729$

If your friend cubes a number between 1 and 9, you will instantly know which number they picked! If they pick a number from 10 to 99, here's what you're going to do. For this example, let's pretend your friend picked **$72^3 = 373{,}248$**.

$$\sqrt[3]{373{,}248}$$

Steps

1. When you cube a number between 10 and 99, it will always be between 1,000 and 999,999. For this first step, you're only going to focus on the digits to the left of the comma (everything left of the last three digits). For example, with 373,248 we will only focus on 373. Now think back to the cubes you memorized earlier. Which two cubes is 373 between? It is between $7^3 = 343$ and $8^3 = 512$. Take the lower number of those, which is 7, and that will be the first digit in your answer!

$5^3 = 125$
$6^3 = 216$
$7^3 = 343$
$8^3 = 512$
$9^3 = 729$

373 248 = 7

2. Next, you will solve the second digit in the answer. For this, focus on the last digit in your number (the 8 in 373,248). Now think back to the cubes you memorized earlier again. Which cube ends in the same number as yours? Both $2^3 = 8$ and 373,248 end in an 8, so the last digit in your answer will be 2! Because every cube ends in a unique number, there should only be one that matches. With these two steps, you can mentally tell your friend that the cube root of 373,248 is 72!

$1^3 = 1$
$2^3 = 8$
$3^3 = 27$
$4^3 = 64$
$5^3 = 125$
$6^3 = 216$
$7^3 = 343$
$8^3 = 512$
$9^3 = 729$

373 248 = 72

Practice

$\sqrt[3]{1{,}331}$ $\sqrt[3]{148{,}877}$ $\sqrt[3]{912{,}673}$

Why Does This Work?

Once you've memorized your cubes for 1^3, 2^3, 3^3, 4^3, 5^3, 6^3, 7^3, 8^3 and 9^3, you've also memorized your cubes for 10^3, 20^3, 30^3, 40^3, 50^3, 60^3, 70^3, 80^3 and 90^3. For example, $\mathbf{2^3 = 8}$ while $\mathbf{20^3 = 8{,}000}$. Similarly, $\mathbf{5^3 = 125}$ while $\mathbf{50^3 = 125{,}000}$. And $\mathbf{8^3 = 512}$ while $\mathbf{80^3 = 512{,}000}$. See the pattern? Cubes for two-digit numbers will have an extra 000 at the end because they are 1,000 times bigger.

Now let's see how we solved our example $\mathbf{373{,}248 = 72^3}$. Would you agree with me that 72^3 is between 70^3 and 80^3? In the same way, 57^3 is between 50^3 and 60^3 and 18^3 is between 10^3 and 20^3. During the first step, we only focused on the 373 in 373,248 and isolated it as 373,000. Because 373,000 is between 343,000 (70^3) and 512,000 (80^3), we know that the first digit must be 7 (the 7 in 72 represents 70 because it is in the tens digit).

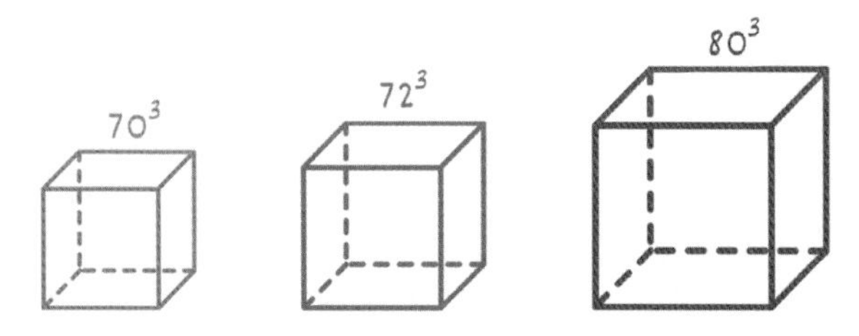

Let's move on to how we did the second step. Why did the ones digit in 373,248 equal the ones digit in our number cubed? Let's use a bit of algebra and represent our two-digit number as $\mathbf{10a+b}$, where "a" is the tens digit and "b" is the ones digit (a = 7 and b = 2 for 72). Now if we cube our number and multiply out the binomials, we get $\mathbf{(10a+b)^3 = 1{,}000a^3 + 300a^2b + 30ab^2 + b^3}$. Notice how all the terms except b^3 have a coefficient that is a power of 10 in front of it. This means that only b^3 will affect the ones digit in our number cubed! Therefore, the ones digit in 373,248 must equal the ones digit in $\mathbf{2^3 = 8}$.

A Mixed Bag of Tricks

Welcome to the grand finale of math tricks! This chapter is a bit of a mixed bag, with tricks that cover a variety of math concepts and real-life scenarios. Get ready to compare fractions in seconds, prove that 1 is the same as 0.999 . . . repeating and master the art of doubling and tripling your money.

And that's not all! We'll also learn how to spot a fake credit card, predict days of the week and explore one of math's most mysterious patterns. The tricks in this chapter may seem unrelated at first, but they all have one thing in common: They are going to make you go "whoa!" and remind you how cool math can be. So, let's dive in and make some math magic!

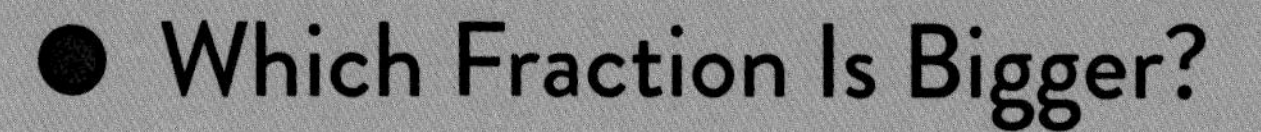

Which Fraction Is Bigger?

Imagine that you are out watching a basketball game.

The home team has an impressive three-point shot success rate of 5 out of 8 tries (⅝) while the visiting team has a solid success rate of 4 out of 7 (4⁄7). So, the question is, which team has the better score when it comes to hitting those long-range shots?

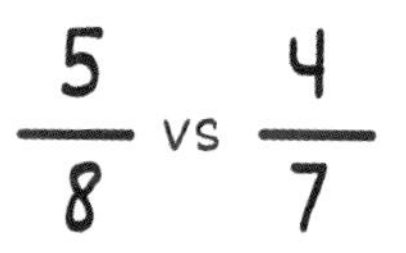

It's not obvious which team is doing better just by looking at ⅝ and 4⁄7, right? Well, I've got a quick and easy trick to help you determine which of these fractions is bigger in under ten seconds!

Steps

1. First, grab the top number of the first fraction (5) and the bottom number of the second fraction (7). Multiply these two numbers (**5 x 7 = 35**) and write 35 on top of the first fraction.

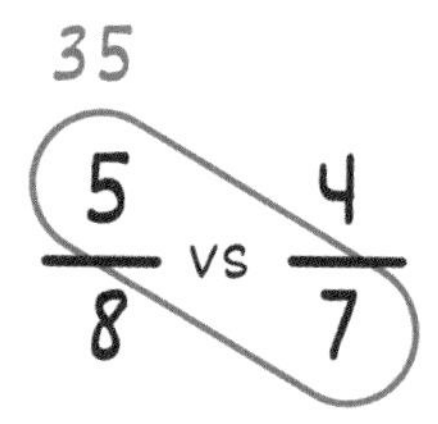

Next, grab the top number of the second fraction (4) and the bottom number of the first fraction (8). Multiply these two numbers (**4 x 8 = 32**) and write 32 on top of the second fraction.

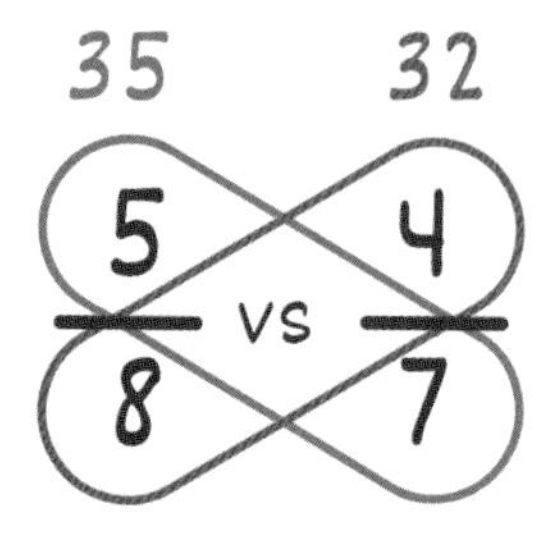

3 Now, which of these numbers that you wrote down (35 vs. 32) is bigger? Clearly 35 is! This means that the fraction underneath the 35 will be bigger. So, ⅝ is bigger than $\frac{4}{7}$ and the home team has a better stat than the visiting team in our example!

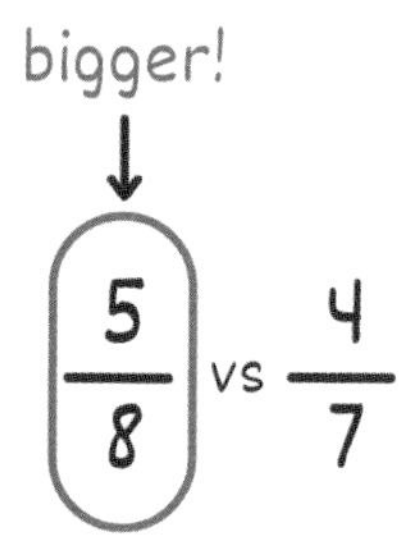

Let's keep the game going! Which fraction is bigger: $\frac{4}{7}$ or $\frac{7}{11}$?

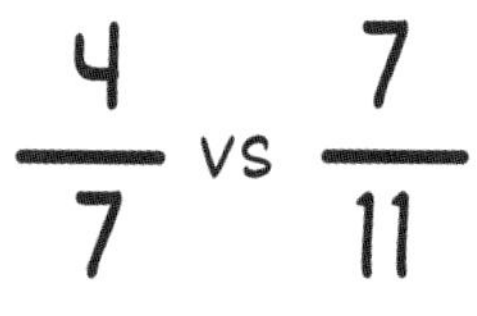

Give it a try and then check out how to do it below.

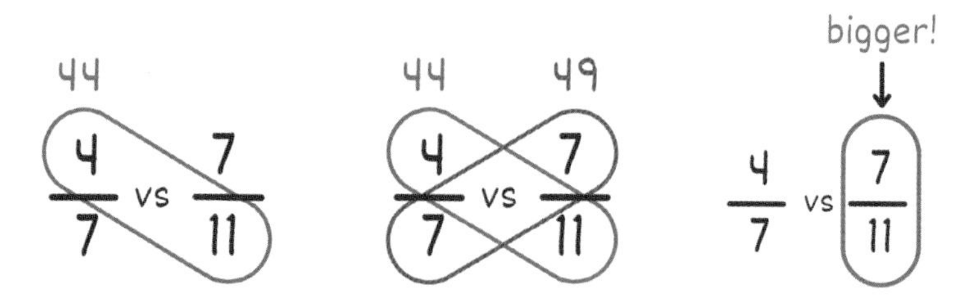

Practice

Which fraction is bigger?

$2/3$ or $7/12$ $4/5$ or $11/13$ $14/20$ or $120/180$

Why Does This Work?

This trick might be a bit of a head-scratcher, but the idea behind it is super simple! Let's start with an example. Can you tell me which of these fractions is bigger: $5/9$ or $8/9$? This one's a no-brainer, right? You knew $8/9$ was bigger just by looking at it because the denominators were the same for both fractions and you could easily compare which numerator was bigger.

That's the key to this trick. Let's go back to the first example.

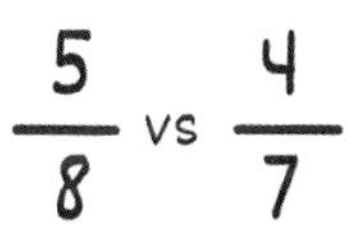

$$\frac{5}{8} \text{ vs } \frac{4}{7}$$

When we're multiplying our numbers, we're also secretly making the denominators for both fractions the same. We don't write the new denominators during the trick, but we're making both fractions out of 56.

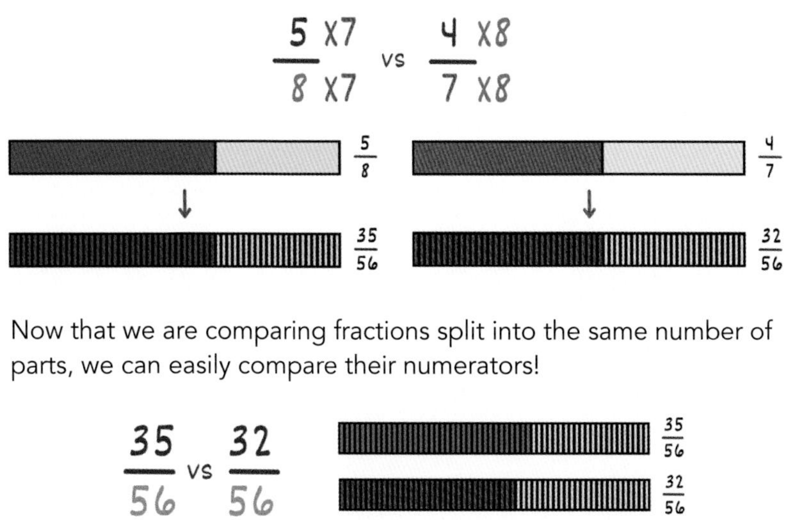

Now that we are comparing fractions split into the same number of parts, we can easily compare their numerators!

 Myth Busted! 1 = 0.999 . . .

Here is something that will leave you questioning everything you thought you knew about math! Did you know that the repeating decimal 0.999 . . . is equal to 1? It may seem impossible, but stick with me here and I'll show you why.

Let's look at the fractions $\frac{1}{3}$ and $\frac{2}{3}$. If we add them together, we get $\frac{1}{3} + \frac{2}{3} = \frac{3}{3} = 1$. But what happens when we add them in decimal form? **0.333 . . . + 0.666 . . . = 0.999 . . .**

$$\frac{1}{3} = 0.33333\ldots$$
$$+\ \frac{2}{3} = 0.66666\ldots$$
$$\frac{3}{3} = 0.99999\ldots$$

And this is just one example—it happens with many other fractions too!

$$\frac{1}{11} = 0.090909\ldots$$
$$+\ \frac{10}{11} = 0.909090\ldots$$
$$\frac{11}{11} = 0.999999\ldots$$

Still not convinced that **0.999 . . . = 1**? Let's prove it with a bit of algebra.

Steps

 Let's say a = 0.999 . . .

$$a = 0.999\ldots$$

2

Let's multiply both sides of the equation by 10. This would give us $10a = 9.999\ldots$

$$a = 0.999\ldots$$
$$10a = 9.999\ldots$$

Here's where things get interesting. Would you agree with me that 9.999 . . . is the same as $9 + 0.999\ldots$?

$$a = 0.999\ldots$$
$$10a = 9.999\ldots$$
$$10a = 9 + 0.999\ldots$$

Because $0.999\ldots = a$, let's replace the 0.999 . . . with "a."

$$a = 0.999\ldots$$
$$10a = 9.999\ldots$$
$$10a = 9 + 0.999\ldots$$
$$10a = 9 + a$$

Let's move every "a" to the left side of the equation. We can do this by subtracting both sides by "a." Our equation becomes $9a = 9$.

$$a = 0.999\ldots$$
$$10a = 9.999\ldots$$
$$10a = 9 + 0.999\ldots$$
$$10a = 9 + a$$
$$-a \qquad -a$$
$$9a = 9$$

Dividing both sides by 9, we end up with $a = 1$. But hold on, at the beginning we said that $a = 0.999\ldots$! Behold, 1 and 0.999 . . . must be the same thing.

$$a = 0.999\ldots$$
$$10a = 9.999\ldots$$
$$10a = 9 + 0.999\ldots$$
$$10a = 9 + a$$
$$-a \qquad -a$$
$$9a = 9$$
$$a = 1$$

Still Doesn't Feel Right?

We just uncovered a mind-blowing revelation that 0.999 . . . is the same as 1 and this is completely true in the world of numbers we live and breathe in. But what if I told you that there's a whole new dimension of numbers out there waiting to be explored? Introducing the mysterious world of hyperreal numbers! This is a place where numbers can be infinitely big and infinitely small, where the impossible becomes possible. In our world, these numbers may seem insignificant, but in the hyperreal world, they are a whole different story.

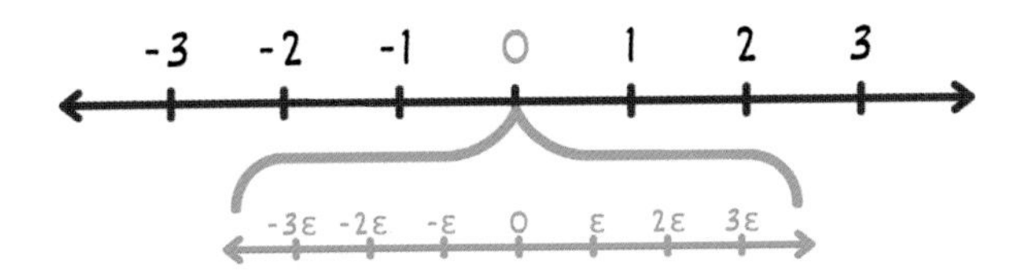

Imagine a world where the tiniest things hold a universe within. The hyperreal world of numbers is such a place, similar to the microscopic realm of atoms you've learned about in chemistry class. Take a look around you: The world you see with your eyes is just the tip of the iceberg. In our world, a grain of sand may seem insignificant, but in the microscopic realm, it holds a whole universe of 50,000,000,000,000,000,000 atoms.

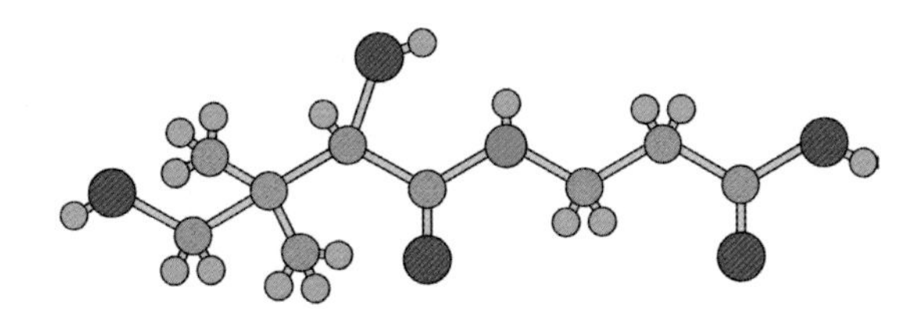

So, in this hyperreal dimension, 0.999 . . . and 1 are their own unique numbers. But in the world of numbers we live in, 0.999 . . . and 1 look identical to us!

Investing 101: How to Double Your Money

Have you ever invested before? One day you may plan on investing your money, and when that day comes, here's a handy trick to keep in mind!

Imagine you're diving into the world of investing and you decide to put your money into a fund like the S&P 500, which has a history of delivering a solid 10% average rate of return per year. That means if you invest $100 today, it'll grow to $110 (**$100 x 110% = $110**) the next year, $121 (**$110 x 110% = $121**) the following year and so on. But have you ever stopped to think about how long it would take to double your initial investment? Don't worry about breaking out the calculator; there's a simple trick to estimate it in your head!

Rule of 72

$$\text{Doubling Time} = \frac{72}{\text{Rate of Return}}$$

Steps

Take 72 and divide it by the rate of return on your investment (also known as the interest rate), and be sure to express the rate of return as a percentage. So, if it's a 10% rate of return, use 10 and not 0.1.

Rule of 72

$$\text{Rate of Return} = 10\%$$

$$\text{Doubling Time} = \frac{72}{10}$$

Solve the problem and you'll know how long it'll take your money to double. In this example, your money would take about 7.2 years to double.

$$\text{Doubling Time} = \frac{72}{10} = 7.2 \text{ years}$$

This handy trick isn't just limited to investments. It can also be used to calculate how long it'll take to pay off debt at a certain interest rate. Imagine you borrow money at an interest rate of 12% but you don't make any payments back. How long will it take for your debt to double? It's simple, just use the trick—72 divided by 12 equals 6 years! So, in 6 years, your debt will have doubled if no payments are made. Pretty eye-opening, isn't it?

Practice

How long will it take you to double your money with a 36% rate of return?

If your money doubled in 12 years, what is your rate of return?

Why Does This Work?

The rule of 72 is a quick and easy approximation to help you estimate the growth of your money based off the exponential growth formula **($FV = PV*(1+r)^t$)** where FV is the future value, PV is the present value, r is the rate and t is the time period. If your goal is to double your money, then **$FV/PV = 2$** and this equation simplifies to **$2 = (1+r)^t$**. Here's a graph of what this looks like—check out how different rates will affect how long it takes to double your money at **$FV/PV = 2$**!

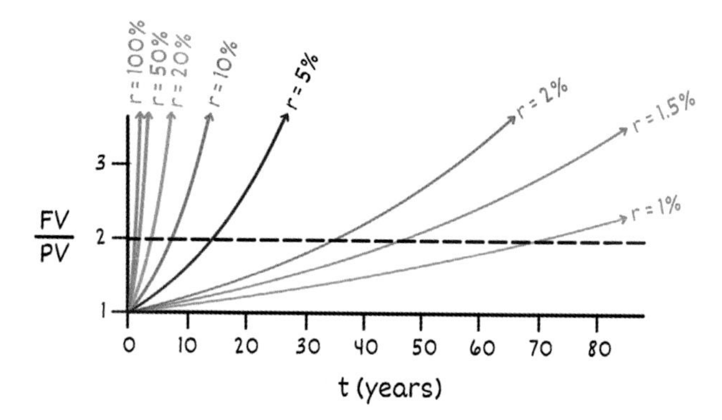

To solve for the "t" we need to take the natural logarithm of both sides (because t is an exponent) to get **$t = \ln(2)/\ln(1+r)$**. Now this is where the approximation comes in—we can approximate **$\ln(1+r) = r$ and $\ln(2) = 0.693$** to get our equation **$t=0.693/r$**. We can then multiply the top and bottom by 100 to get **$t=69.3/r$** where r is expressed as a percentage. But hold on! Why did we get 69.3 instead of 72? Well, because 72 easily divides into many common rates (like 1, 2, 3, 4, 6, 8, 9 and 12), it is much easier to divide 72 instead of 69.3 in our heads, so using 72 has gained popularity as a value to estimate the doubling time. This is an approximation, after all!

Forget Doubling, Let's Triple Your Money!

You've unlocked the secret to doubling your money, but why stop there? Let's shoot for the moon and learn how to triple it!

Just like before, let's say you invest your money in the stock market with an average rate of return of 10% per year. Using the rule of 72, we know it would take 7.2 years to double your investment. But what if I told you there's also a quick way to estimate the time it takes to triple your investment? Meet the rule of 115!

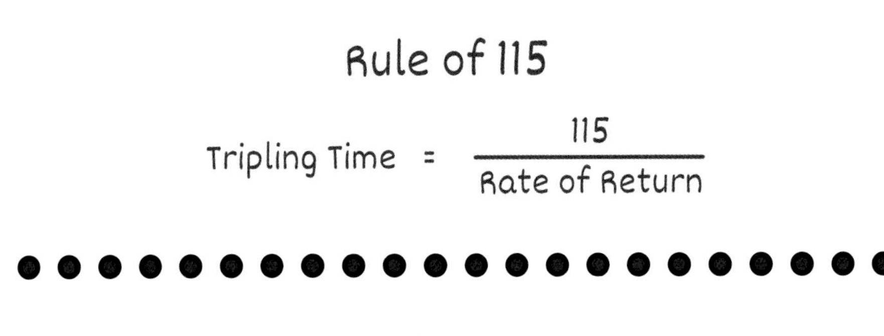

Steps

Take the magic number 115 and divide it by the rate of return on your investment. Be sure to express the rate of return as a percentage (10% instead of 0.1).

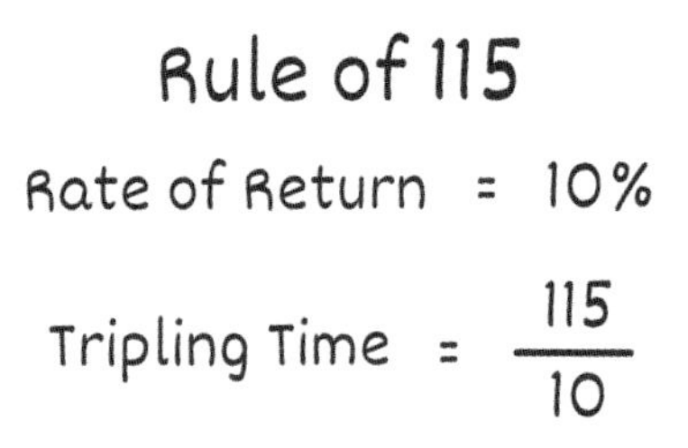

Calculate this out to get your answer! In this example, your money would take 11.5 years to triple.

$$\text{Tripling Time} = \frac{115}{10} = 11.5 \text{ years}$$

Here's something to think about throughout life. With a 10% rate of return on your investment each year, it took a whole 7.2 years for your money to double, but only 4.3 years more (11.5 years total) for your money to triple! And the excitement doesn't end here—in just 3 more years, your money quadruples and in a mere 2.3 more years, it quintuples. The amount of time it takes to grow gets less and less each year. This is the power of the *compound effect* and it's a huge reason why you should start saving and investing your money as early as possible. So, try it out for yourself and watch your wealth grow exponentially!

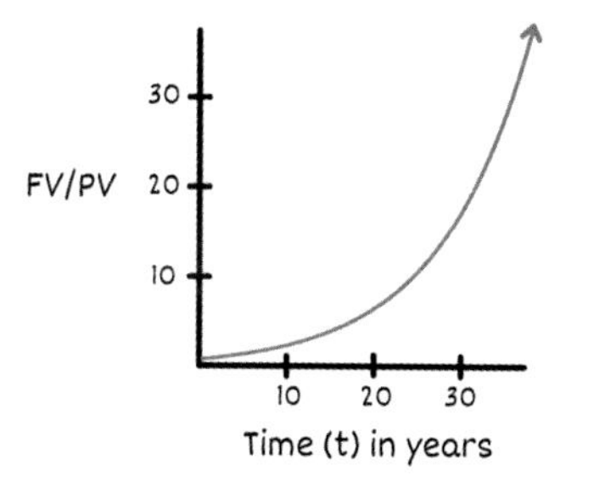

Practice

How long will it take you to triple your money with a 23% rate of return?

If your money doubled in 14.4 years, how many years would it take to triple?

Why Does This Work?

The rule of 115 is a powerful tool that can help you reach your financial goals! Just like the rule of 72, it is based on the exponential growth formula **($FV = PV*(1+r)^t$)** where FV is the future value, PV is the present value, r is the rate and t is the time period. When you triple your money, **$FV/PV = 3$** and this equation simplifies to **$3 = (1+r)^t$**. Take a glance at this graph; it shows you how your money would grow with different rates! The larger the rate, the faster it will take your money to triple at **$FV/PV = 3$**!

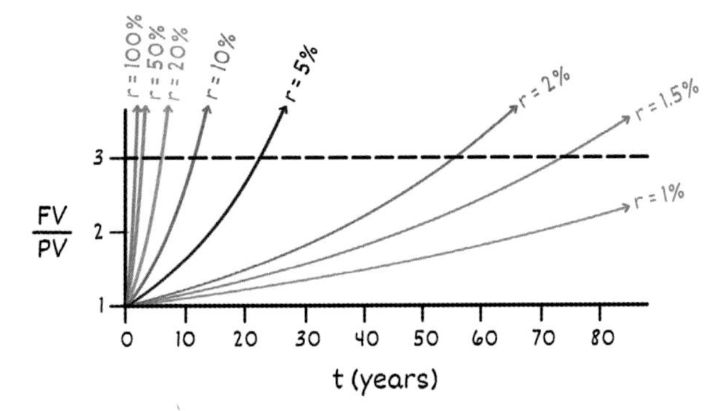

To solve for the time "t," let's take the natural logarithm of both sides to get **$t = \ln(3)/\ln(1+r)$**. This is where we can start using some approximation magic—we can approximate **$\ln(1+r) = r$** and **$\ln(3) = 1.0986$** to get our equation **$t=1.0986/r$**. We can then multiply the top and bottom by 100 to get **$t=109.86/r$**, where r is expressed as a percentage. But you may be wondering, why did we get 109.86 instead of 115? Well, the answer is simple—115 can be divided easily into many numbers, making it a popular choice among investors as an approximation!

The Time Traveler's Trick

Are you ready to take a journey through time? Imagine being able to impress your friends by asking them to pick any date in the past or future and then telling them which day of the week it falls on. The secret? It's all in the charts!

To pull off this incredible trick, all you need to do is commit a few charts to memory. First up, let's talk about the days of the week.

Days of the Week

Sunday	Monday	Tuesday	Wednesday	Thursday	Friday	Saturday
0	1	2	3	4	5	6

Each day is assigned a number, making it easy to remember. Sunday is 0, Monday is 1 and so on, all the way to Saturday being 6. And that's all you'll need to know for now; we'll come back to this chart later.

Next up we have the month chart.

Month Code

Jan	Feb	Mar	Apr	May	Jun	Jul	Aug	Sep	Oct	Nov	Dec
0	3	3	6	1	4	6	2	5	0	3	5

You might be thinking, "Oh no, I have to remember twelve numbers," but don't worry, it's not as difficult as it seems. I like to break down the twelve numbers into groups of three: 033, 614, 625, 035. Can you spot any patterns? The first and last groups both begin with 03 and the second and third groups both start with 6. Plus, the last two digits of the third group (2 and 5) are just a step up from those in the second group (1 and 4).

And now, let's tackle the century codes.

Century Code

1600 - 1699	1700 - 1799	1800 - 1899	1900 - 1699	2000 - 2099	2100 - 2199	2200 - 2299	2300 - 2399
6	4	2	0	6	4	2	0

It's simple, I promise! Just remember the numbers 6, 4, 2, 0. This pattern keeps repeating before 1600 and after 2399 as well.

With these three charts in your back pocket, you're all set to master this trick. So, when you're ready, take a deep breath and let's go through the steps together, nice and slowly.

Let's say you want to find the day of the week for March 5, 2009.

Steps

1. There are two parts to this trick. In the first part, we'll add up five crucial numbers. And in the second part, we'll use that sum to get the day of the week.

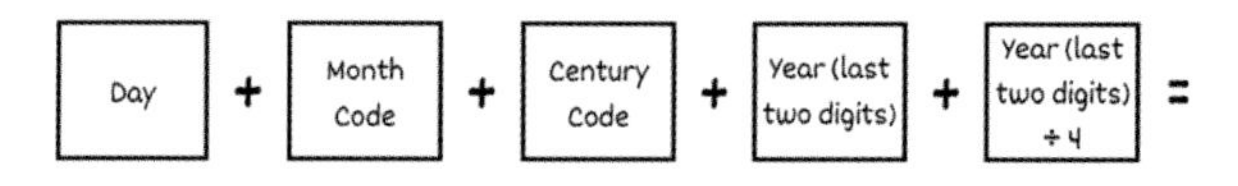

The first number you'll add is the day. For March 5, the day will be 5.

5 + Month Code + Century Code + Year (last two digits) + Year (last two digits) ÷ 4 =

Now, let's move on to adding the month code. Using our handy chart, we see that the code for March is 3.

Jan	Feb	Mar	Apr	May	Jun	Jul	Aug	Sep	Oct	Nov	Dec
0	3	3	6	1	4	6	2	5	0	3	5

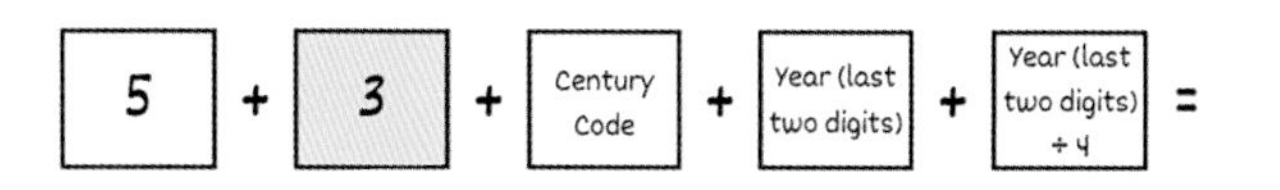

Next up, we'll add the century code. Because 2009 falls between 2000 to 2099, the century code will be 6.

1600 - 1699	1700 - 1799	1800 - 1899	1900 - 1699	2000 - 2099	2100 - 2199	2200 - 2299	2300 - 2399
6	4	2	0	6	4	2	0

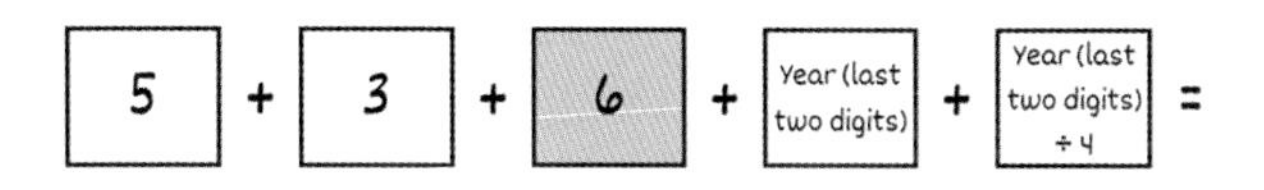

Next, add the last two digits of the year. The last two digits in 2009 are 09, so add a 9 in your box!

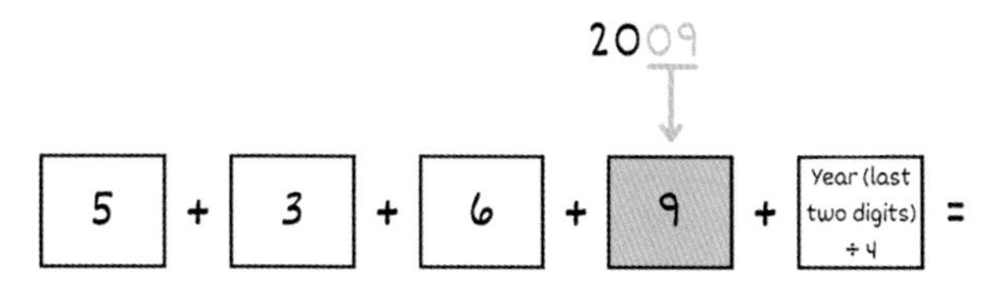

For the last number, let's take that number we just added (9) and have some fun by dividing it by 4 (**9 ÷ 4 = 2 R1**). Write the quotient (2) in your last box. Don't fret if there's a remainder, just ignore it and move on to the next step!

5 + 3 + 6 + 9 + 2 =

9 ÷ 4 = 2 R1

7. Finally, sum up the five crucial numbers you wrote down: **5 + 3 + 6 + 9 + 2 = 25**.

$$5 + 3 + 6 + 9 + 2 = 25$$

8. Now the moment we've all been waiting for: Let's find out what day of the week March 5, 2009, was on! Let's take that sum we've been working on (25) and divide it by 7 (**25 ÷ 7 = 3 R4**). This time we are going to ignore the quotient and only pay attention to the remainder. Take the remainder (4) and match it to the day of the week, and ta-da! It was a Thursday! So, March 5, 2009, was a Thursday.

$$= 25 \div 7 = 3\ R4$$

Sunday	Monday	Tuesday	Wednesday	Thursday	Friday	Saturday
0	1	2	3	4	5	6

Are you ready to try one on your own? Let's find out what day of the week October 9, 2215, will be. Take your time to use the charts and check back for the answer. The day code is 9, the month code is 0, the century code is 2, the last two digits of the year are 15 and the quotient of the last two digits of the year divided by 4 is 3. Let's add all these numbers together and get **9 + 0 + 2 + 15 + 3 = 29** and then divide **29 ÷ 7 = 4 R1**. Now, it's time to match the remainder of 1 with the day of the week. Are you ready for the answer? It's going to be a Monday!

$$9 + 0 + 2 + 15 + 3 = 29$$

$$= 29 \div 7 = 4\ R1$$

Sunday	Monday	Tuesday	Wednesday	Thursday	Friday	Saturday
0	1	2	3	4	5	6

Leap Year Exceptions

Now that you've gotten the hang of this, I want to tell you about an exception to this trick—leap years! As you might know, leap years occur every four years, but when your date falls on both a leap year and January/February, you'll need to make a small adjustment and subtract 1 from the final sum. For example, let's take a look at January 3, 2024. To solve this, we add everything up as usual (**3 + 0 + 6 + 24 + 6 = 39**) and then divide our sum by 7 (**39 ÷ 7 = 5 R4**), but instead of matching the remainder of 4 to the day of the week, we need to first subtract 1 from it (**4 - 1 = 3**) and then match 3 with the day of the week, which is Wednesday!

3 + 0 + 6 + 24 + 6 = 39

= 39 ÷ 7 = 5 R4

= 4 - 1 = 3

Sunday	Monday	Tuesday	Wednesday	Thursday	Friday	Saturday
0	1	2	3	4	5	6

How Do You Know Which Years Are Leap Years?

A leap year occurs every four years, and the way to know whether the year is a leap year is to check whether the last two digits of the year are divisible by 4. For example, 2032 is a leap year, because 32 is divisible by 4 (**32 ÷ 4 = 8**) and 1924 is a leap year (**24 ÷ 4 = 6**) but 2022 is not a leap year because 22 is not divisible by 4. The only exception is that the years that end in two zeros but are not divisible by 400 are not leap years (for example, 1600, 2000 and 2400 are leap years but 1700, 1800, 1900, 2100, 2200, 2300 and 2500 are not)!

But wait, before you start traveling through time, there's one more thing to keep in mind. This trick works for any date as long as it's after the year 1582. You may be wondering why. The calendar we all know and love is formally known as the Gregorian calendar and it was only introduced in October 1582. Before that, there were no such thing as leap years. Because the Earth rotates around the sun every 365.25 days, a year is actually 365.25, not 365, days! To fix this, leap years were introduced to keep the calendar from drifting.

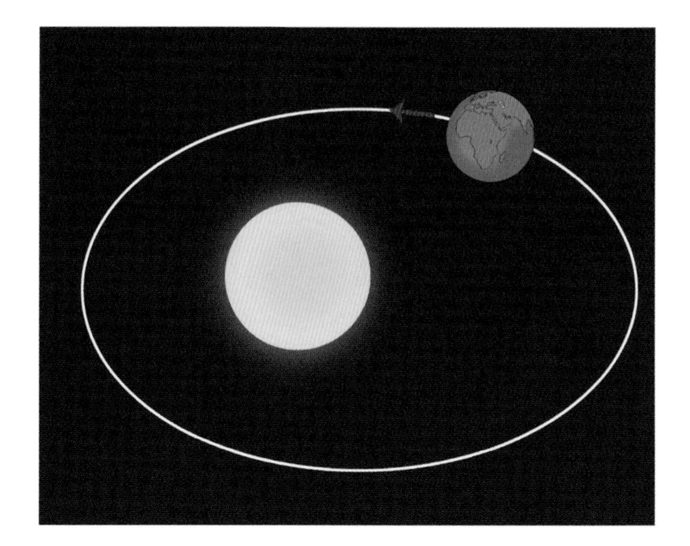

Now that you know the trick, you can impress your friends and family by telling them what day of the week their birthday or any other special date falls on. You can even use this trick to plan ahead for future events and make sure you're always on time. Happy time traveling!

Practice

Can you get the day of the week for these famous dates in history?

The Declaration of Independence (July 4, 1776)

Albert Einstein was born on Pi Day (March 14, 1879)

Neil Armstrong and Buzz Aldrin landed on the moon (July 20, 1969)

How to Spot a Fake Credit Card

Are you ready to become a secret detective and spot fake credit cards? This trick is not known by many, but it is something that happens behind the scenes in your life every day. So put on your detective hat and let's crack this case!

Have you ever looked closely at a credit card? All of them have a long string of numbers, sixteen digits to be exact. These numbers may look random, but they are far from it. How would you distinguish the sixteen digits of a real credit card from just any random set of sixteen numbers? Here's the secret!

Steps

1. First, grab your pen and paper and jot down the sixteen digits of your credit card.

   4100 2328 8521 1367

2. Next, starting with the first digit, take every other digit and multiply it by 2.

   4100 2328 8521 1367

   x2 ↓ ↓ ↓ ↓ ↓ ↓ ↓ ↓

   8 0 4 4 16 4 2 12

If any of these new numbers are 10 or above, break them down into single digits by adding the digits together (16 becomes 7 from 1 + 6 and 12 becomes 3 from 1 + 2).

4100 2328 8521 1367

8 0 4 4 16 4 2 12

1+6 = 7 1+2 = 3

Now, slot these new digits back into your original sixteen-digit number.

8100 4348 7541 2337

Now for the moment of truth! Let's add all sixteen digits together. If the sum is not a multiple of 10 (like 10, 20, 30, 40, 50, 60, 70, 80 and so on), then your card is a fake! Because the numbers in this example's card add up to 60, this card is most likely real!

8+1+0+0+4+3+4+8+7+5+4+1+2+3+3+7 = 60

Imagine you're shopping for the perfect pair of shoes online, and in the midst of entering your credit card information, you accidentally switch the second and third digits (4010 2328 8521 1367 instead of 4100 2328 8521 1367). But don't worry, the payment system will run your credit card number through the steps we just went through and in a few seconds alert you that you typed it in incorrectly!

Practice

Are these credit card numbers real or fake?

4245 3102 6713 1134

3421 1589 4001 3897

5133 4857 4363 1949

Why Does This Work?

This process is called the *Luhn Algorithm*, also known as the *Modulus 10* or *Mod 10 Algorithm*. This simple process is a verifier that helps distinguish valid numbers from those that have been mistyped or entered incorrectly and is now used by most credit card companies and governments.

But what makes it so effective? By doubling every alternate number and taking the sum of all the digits, this process can easily spot any typos or errors if the sum does not equal to a multiple of 10. So next time you use a credit card, try out the Luhn Algorithm and double-check your numbers!

The 6174 Mystery

Are you prepared to unlock the secrets of the unknown? Allow me to introduce you to the mysterious number 6174, also known as *Kaprekar's constant*. All you must do is pick a four-digit number, follow these steps and let the mystery unveil itself to you. Beware, for the truth may change your perception of reality forever!

7283

Steps

1. Let's play with our numbers. Take a four-digit number (say, 7283) and get ready to rearrange its digits. First, you'll arrange them in descending order from the biggest to smallest digit—8732. Then, you'll switch gears and arrange them in ascending order from the smallest to the biggest digit—2378.

7283

Descending = 8732

Ascending = 2378

2. Next, simply subtract the smaller number from the bigger number.

7283

8732 - 2378 = 6354

3. Keep repeating these steps until you reach the ultimate destination: 6174. Once you reach 6174, no matter how many times you repeat the process you will forever remain at 6174. For example, if we arrange 6174 in descending order (7641), then ascending order (1467) and then subtract the two, we would end right back at 6174 **(7641 - 1467 = 6174)**.

7283

8732 - 2378 = 6354

6543 - 3456 = 3087

8730 - 0378 = 8352

8532 - 2358 = 6174

4. All right, what was so special about that? You see, the beauty of this process is that you can pick any four-digit number and within seven iterations of steps 1 through 3, you will always end up at the mystical number 6174. Let me show you a few more examples to prove it. Some numbers may reach 6174 in just two steps, while others may take the full seven!

9990

9990 - 0999 = 8991

9981 - 1899 = 8082

8820 - 0288 = 8532

8532 - 2358 = 6174

3735

7533 - 3357 = 4176

7641 - 1467 = 6174

1042

4210 - 0124 = 4086

8640 - 0468 = 8172

8721 - 1278 = 7443

7443 - 3447 = 3996

9963 - 3699 = 6264

6642 - 2466 = 4176

7641 - 1467 = 6174

Eerie, isn't it? Now it's your turn! Pick a random four-digit number and give this a try. Just a heads up—you can pick any four-digit number (even ones with leading zeros like 0028 or trailing zeros like 8000); however, avoid the nine numbers with repeating digits (1111, 2222, 3333, 4444, 5555, 6666, 7777, 8888 and 9999). Those are the only four-digit numbers that will not converge to 6174.

Why Does This Work?

The unique pattern of Kaprekar's constant remains a mystery until this date, but thanks to the power of coding, we can now take a closer look at this pattern. We can see that all four digit numbers that follow this process (subtracting the ascending-ordered digits by its descending-ordered digits) converge at 6174 following one of several paths. Want to see it for yourself? Pick a random number and watch as it follows one of these paths to 6174!

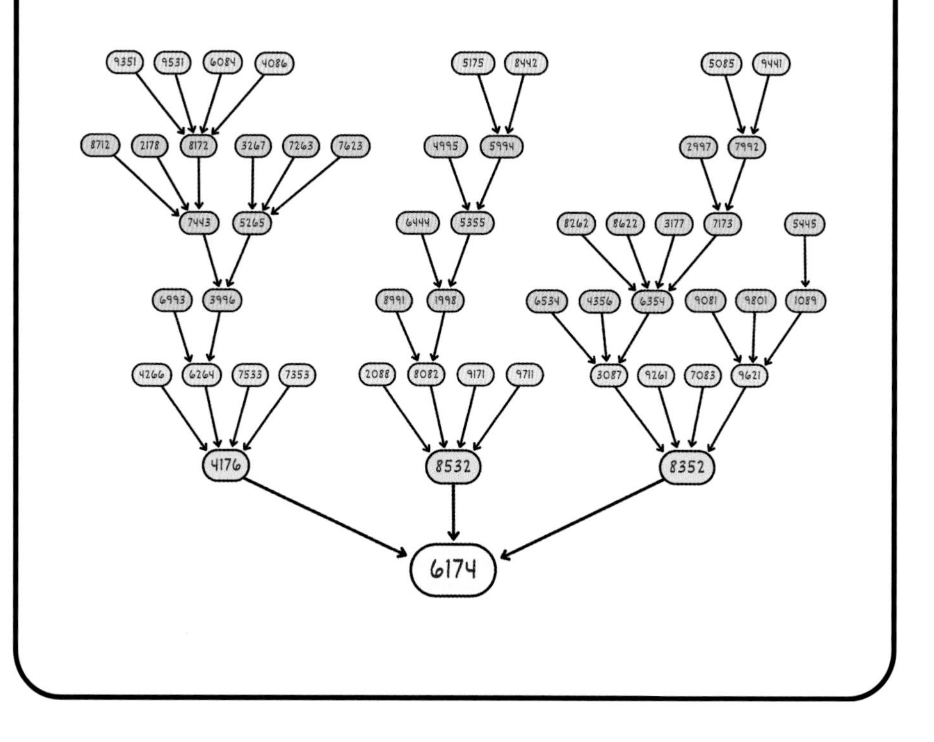

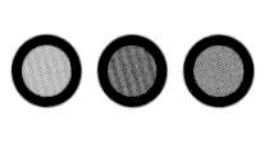

But that's not all! When we map all the numbers from 0 to 10,000 on a 100 by 100 grid and color code them by the number of iterations it takes to convert to 6174, we come up with a beautiful pattern that is like a piece of art. This is the mysterious and beautiful world of numbers!

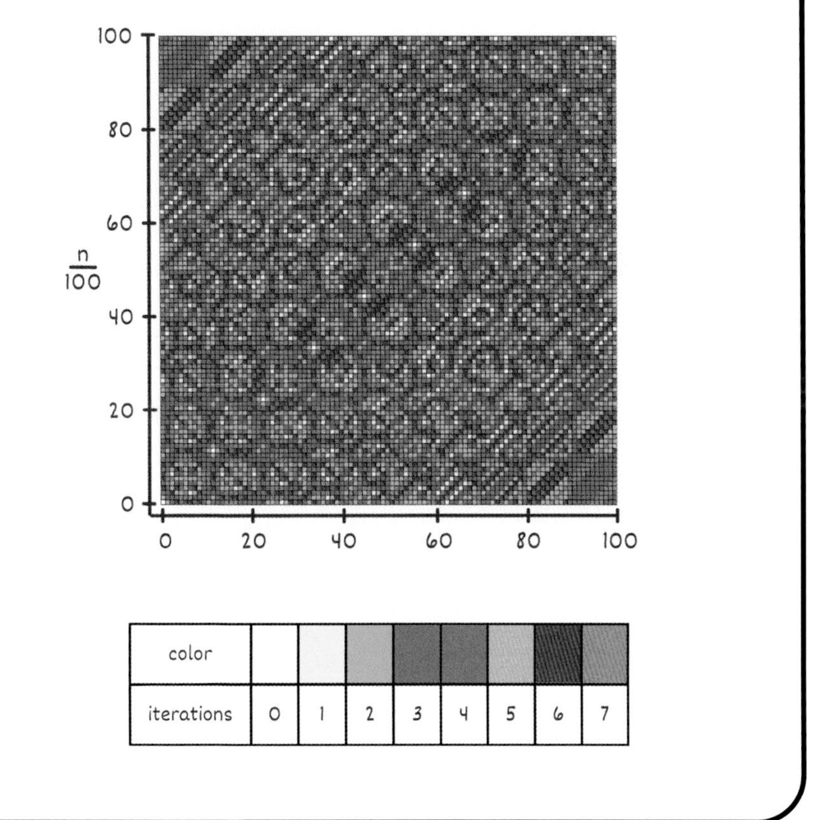

color								
iterations	0	1	2	3	4	5	6	7

Answers

Add Odd Numbers in Seconds

$1 + 3 + 5 + 7 + 9 + 11 + 13 + 15 + 17 + 19 + 21 = 11^2 = 121$

Sum of odd numbers from 1 to 199 $= 100^2 = 10{,}000$

Sum of all odd numbers up to 2,007 $= 1{,}004^2 = 1{,}008{,}016$

Add Even Numbers Mentally

$2 + 4 + 6 + 8 + 10 + 12 + 14 + 16 + 18 = 9 \times (9 + 1) = 90$

Sum of even numbers from 1 to 20 $= 10 \times 11 = 110$

Sum of all even numbers up to $1{,}000 = 500 \times 501 = 250{,}500$

Subtract Big Numbers without Borrowing

$700 - 83 = 617$

$17{,}000 - 936 = 16{,}064$

$-238 + 5{,}000 = 4{,}762$

Subtract by Adding?!

$52 - 17 = 35$

$1{,}234 - 321 = 913$

$3{,}920 - 1{,}242 = 2{,}678$

Multiplying by 5? Do This!

$27 \div 2 = 13.5$, $13.5 \times 10 = 135$

$120 \times 5 = 120 \div 2 \times 10 = 600$

$64 \times 5 = 64 \div 2 \times 10 = 320$

This 11 Trick Will Blow Your Mind

$53 \times 11 = 583$

$86 \times 11 = 946$

$7{,}253 \times 11 = 79{,}783$

Multiply Teen Numbers in Your Sleep

$14 \times 15 = 210$

$13 \times 18 = 234$

$17 \times 19 = 323$

Two Apart? Too Easy!

$13 \times 11 = 12^2 - 1 = 143$

$79 \times 81 = 80^2 - 1 = 6{,}399$

$301 \times 299 = 300^2 - 1 = 89{,}999$

Two-Digit Multiplication Rainbows

$13 \times 21 = 273$

$42 \times 14 = 588$

$95 \times 72 = 6{,}840$

Easy Three-Digit Multiplication

121 × 31 = 3,751

821 × 23 = 18,883

458 × 72 = 32,976

Big Numbers? Count Balloons!

How many zeros will the answer for 20 × 10,100 have? Answer: 3 trailing zeros

10,300 × 20 = 206,000

2,000 × 40 × 700 = 56,000,000

When in Doubt, Draw a Box

Draw the box for 362 × 12,803

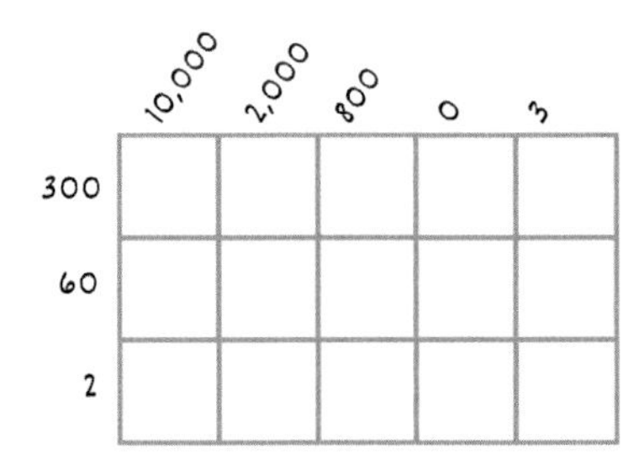

27 × 8,130 = 219,510

123 × 123 = 15,129

Tired of Multiplying? Use Lines!

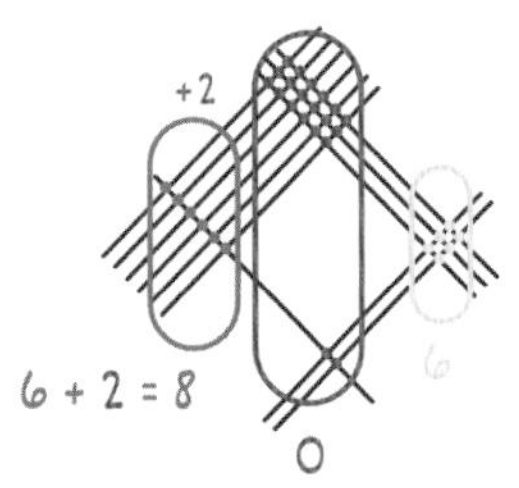

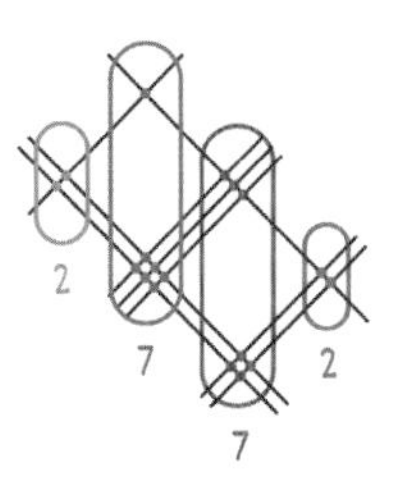

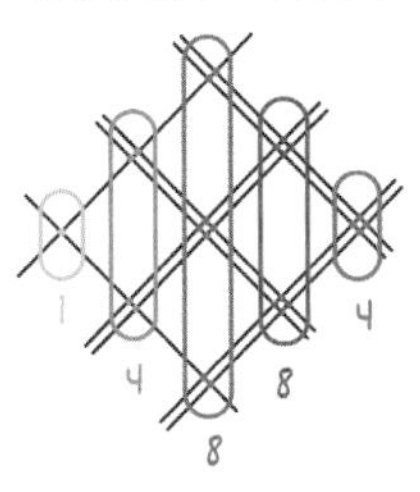

Division? No Sweat!

Here's another way to cut a cake into 8 equal pieces. First, make 2 cuts from the top.

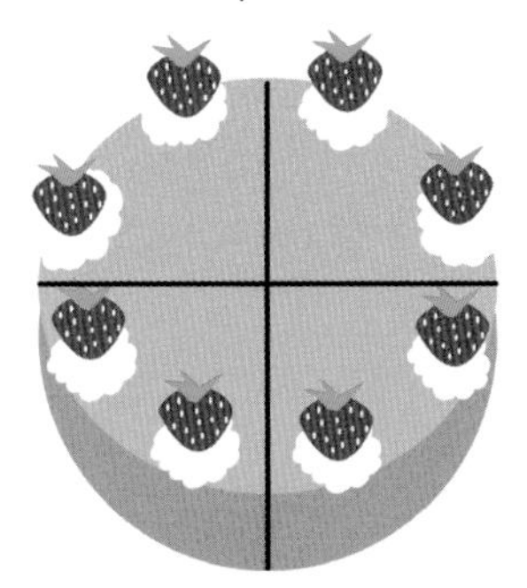

This will create 4 pieces. Then, cut through the middle of the cake on its side. Now you'll have 8 equal pieces!

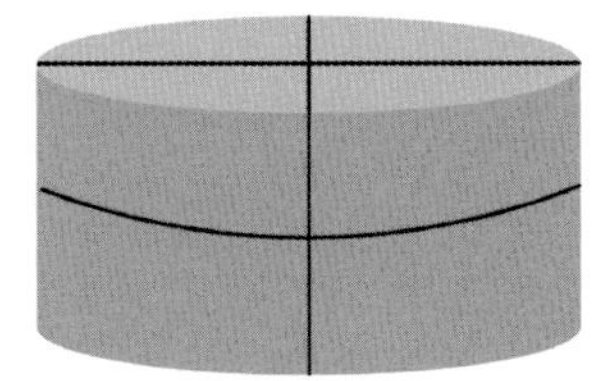

Instantly Divide by 5 (and 0.5, 50, 500)

32 ÷ 5 = 32 × 2 ÷ 10 = 6.4

231 ÷ 50 = 231 × 2 ÷ 100 = 4.62

4,200 ÷ 500 = 4,200 × 2 ÷ 1,000 = 8.4

Can You Divide by 25 (0.25, 2.5, 250)?

112 ÷ 25 = 112 × 4 ÷ 100 = 4.48

1 ÷ 2.5 = 1 × 4 ÷ 10 = 0.4

90 ÷ 250 = 90 × 4 ÷ 1,000 = 0.36

Divide by 125 (0.125, 12.5, 125) in Your Head

100 ÷ 125 = 100 × 8 ÷ 1,000 = 0.8

25 ÷ 0.125 = 25 × 8 ÷ 1 = 200

30 ÷ 12.5 = 30 × 8 ÷ 100 = 2.4

Don't Love Long Division? Try This!

82 ÷ 15 = 5 R7

723 ÷ 80 = 9 R3

850 ÷ 110 = 7 R80

Predict Whether a Number Is Divisible from 2 to 10

78 is divisible by 2, 3 and 6

864 is divisible by 2, 3, 4, 6, 8 and 9

5,040 is divisible by 2, 3, 4, 5, 6, 7, 8, 9 and 10

Stuck? Reverse Your Percentages

23% of 200 = 200% of 23 = 2 × 23 = 46

15% of 20 = 20% of 15 = 15 ÷ 5 = 3

12% of 75 = 75% of 12 = ¾ × 12 = 9

Percentages Made Easy!

90% of 20 = 9 × 2 = 18

30% of 500 = 3 × 5 × 10 = 150

80% of 8 = 8 × 8 ÷ 10 = 6.4

Chop Your Percentages

25% of 48 = 12 (try 25% = 10% + 10% + 5%)

31% of 60 = 18.6 (try 31% = 10% + 10% + 10% + 1%)

19% of 50 = 9.5 (try 19% = 20% - 1%)

Of-Is-What Magic

What is 40% of 75 . . . ? = 40% × 75 . . . ? = 30

20 is what percentage of 50 . . . 20 = ?% × 50 . . . ? = 40%

70% of 20 is what number . . . 70% × 20 = ? . . . ? = 14

Square Numbers from 1 to 99 in Your Head

$13^2 = 169$

$32^2 = 1,024$

$72^2 = 5,184$

Become a Squaring Master (100 to 999)

$111^2 = 12{,}321$

$132^2 = 17{,}424$

$541^2 = 292{,}681$

$15^2, 25^2, 35^2, 45^2, 55^2, 65^2, 75^2, 85^2, 95^2$ in 3 Seconds!

$25^2 = 625$

$55^2 = 3{,}025$

$95^2 = 9{,}025$

The Trick to Cubing Numbers

$11^3 = 1{,}331$

$17^3 = 4{,}913$

$32^3 = 32{,}768$

Become a Square Root Calculating Machine!

$\sqrt{10} \approx 3\frac{1}{6} \approx 3.167$

$\sqrt{52} \approx 7\frac{3}{14} \approx 7.214$

$\sqrt{93} \approx 9\frac{12}{18} \approx 9\frac{2}{3} \approx 9.667$

Mentally Calculate Huge Roots like √6,724

$\sqrt{169} = 13$

$\sqrt{1{,}521} = 39$

$\sqrt{9{,}216} = 96$

Cube Root Magic: From 1 to 999,999

Cube root of 1,331 = 11

Cube root of 148,877 = 53

Cube root of 912,673 = 97

Which Fraction Is Bigger?

$\frac{2}{3}$ or $\frac{7}{12}$ = $\frac{2}{3}$

$\frac{4}{5}$ or $\frac{11}{13}$ = $\frac{11}{13}$

$\frac{14}{20}$ or $\frac{120}{180}$ = $\frac{14}{20}$

Investing 101: How to Double Your Money

How long will it take you to double your money with a 36% rate of return? = 72 ÷ 36 = 2 years

If your money doubled in 12 years, what is your rate of return? = 72 ÷ 12 years = 6%

Forget Doubling, Let's Triple Your Money!

How long will it take you to triple your money with a 23% rate of return? = 115 ÷ 23 = 5 years

If your money doubled in 14.4 years, how many years would it take to triple?

1. First find the rate of return = 72 ÷ 14.4 years = 5% rate of return
2. Then calculate the number of years it would take to triple = 115 ÷ 5 = 23 years

The Time Traveler's Trick

The Declaration of Independence (July 4th, 1776) = Thursday

Albert Einstein was born on Pi Day (March 14, 1879) = Friday

Neil Armstrong and Buzz Aldrin landed on the moon (July 20, 1969) = Sunday

How to Spot a Fake Credit Card

4245 3102 6713 1134 = 60 = Real

3421 1589 4001 3897 = 77 = Fake

5133 4857 4363 1949 = 80 = Real

Acknowledgments

I could not have written this book without some amazing people scattered around the globe!

First and foremost, a big thank you to my fiancé, Ming, for being my rock and giving me the courage to pursue a career I found meaning in. I am so excited to embark on life's journey together with you . . . rocket on! Let's not forget that the mastermind behind the name Pink Pencil Math is none other than Ming!

Next, I couldn't have done this without Pink Pencil Math's #1 fan and #1 stalker, my dad, Paul. I know it must have been a surprise when I left my engineering job to start a math TikTok site, but you stuck it out with me and supported me each step of the way. Thank you for everything.

Following my dad, we have my mom, Fiona. I'd like to believe that the carrot juice you made me drink every morning for ten years contributed more to this book than turning my skin orange. Thank you for your daily smiles and warm hugs. Your infectious positivity inspires me to spread happiness and kindness to others every day.

Next, we have my sister, Amanda. Your text messages filled with emojis never fail to brighten up my day, especially during times when I feel stuck. Your unwavering support and encouragement mean the world to me, and I am grateful for your love. I can't wait to see all the amazing things you will accomplish in your life!

A big thank you goes out to all my friends, who have been incredibly supportive and positive throughout my journey. From our late-night conversations, to your valuable advice when I was feeling lost, to celebrating each win along the way and to all the laughs that have come out of this, you all have made my journey with Pink Pencil Math truly special. Love you all!

Thank you to all those who helped me review the math in this book: Ivana Lee, Arlene Resendiz, Luqman Rahamat and Rizwan Maqsood.

Thanks to Franny and the Page Street team for all the hard work they've poured into this book and their support throughout the entire process. Thank you for believing in me to bring this project to life!

And lastly, my heartfelt gratitude goes out to all my content viewers and supporters. Your enthusiastic engagement and encouragement have been a driving force behind my passion for teaching mathematics and give me the strength to keep moving forward. I couldn't have done any of this without you, and I am continually grateful for all that you have taught me on this journey!

About the Author

Tanya Zakowich is the creator of @pinkpencilmath, where she shares tips and tricks on solving math problems.

To date, her bite-sized videos have amassed hundreds of millions of views, and her Math Foundations course has helped thousands of learners regain their confidence in the subject.

Before @pinkpencilmath, Tanya studied mechanical engineering at Columbia University and worked at NASA, Boeing and Hyperloop One.

In her spare time, Tanya loves planning road trips, fishing with her dad and trying tempura restaurants.

You can find her at @pinkpencilmath on social media or www.pinkpencilmath.com.

Index

V

W